THE STORY OF THE FORD GRAND PRIX ENGINE
Its Design and Development

THE STORY OF THE FORD GRAND PRIX ENGINE

Its Design and Development

by

John Blunsden and David Phipps

1971

ROBERT BENTLEY, INC.

872 Massachusetts Avenue

Cambridge Massachusetts, 02139

Library of Congress Catalog
Card Number 74-163124
ISBN 0-8376-0052-9

Manufactured in Great Britain

Chapters

Introduction 8

1 THE LESSON OF 1961 11

Controversial 1½ litre formula; 1961: Ferrari domination; 1962: Championship for BRM and Hill; 1963: Seven GPs for Clark and Lotus-Climax; 1964: Ferrari and Surtees stage comeback; 1965: Second championship for Clark and Lotus-Climax; Coventry Climax shock announcement; Preparations for 3 litre formula.

2 ANY OTHER BUSINESS 22

Walter Hayes joins Ford; Competitions and Public Affairs; Ford's performance image; Cortina in competitions; Ford's motorsport philosophy; Ford in FJ and F3; Cosworth Ford SCA F2 engine; Colin Chapman's search for GP engine; Ford sanction FVA F2 and DFV F1 engines; Agreement between Ford, Cosworth and Lotus.

3 FOUR REMARKABLE MEN 36

Career and personal profiles on Keith Duckworth, Mike Costin, Bill Brown and Benny Rood; Growth of Cosworth Engineering.

4 DUCKWORTH'S DESIGN AND PHILOSOPHY 51

Brief specification of DFV engine; Importance of cylinder head; dangers of calculating loads; SCA Heron head; Reasons for V8 configuration; Four valves per cylinder; Evaluation of rival engines; Alternative valve layouts; Designing from scratch; Gas flow and combustion; Bore and stroke; Engine as stressed chassis member; Mounting of ancillaries; Mechanical efficiency; Vibration problems; Mechanical losses; Crankcase design; Metal-to-metal joints; Flywheel and starter motor; Choice of materials; Fuel injection and ignition; Scheming the layout; Drawing and detailing; Alternative solutions; Importance of designing simply.

5 1966: THE RETURN OF POWER 64

Jack Brabham proves V8 concept; 3 litre Repco; Review of 1966 races; Cooper-Maserati V12; Ferrari use V6 and V12; BRM H16; Lotus interim models; McLaren's first GP car; Honda and Eagle-Weslake V12s.

6 1967: V8 FOR VICTORY 80

Lotus 49 initial tests; Delayed debut of DFV engine; Early 1967 races reviewed; Modifications to rival engines; Matra enter F1; Zandvoort and first-time victory for Lotus-Ford.

7 A MESSAGE FROM THE SPONSOR 98

Remaining 1967 races reviewed; Another Repco Brabham championship; Clark and Hill suffer setbacks in Europe; Ford engine for other teams; Drama and success at Watkins Glen.

8 1968: CHAMPIONS OF THE WORLD 120

Year's GP races reviewed; Matra and McLaren use Ford power; 2½ litre DFW engine; Ford sports prototype; death of Jim Clark; Ill-fated V8 Honda; BRM turn to V12; Failure of four-camshaft Repco; Emergence of wings; Hill and Lotus-Ford win championship.

9 1969: ENTENTE CORDIALE 147

Year's GP races reviewed; Brabham uses Ford engine; Cooper and Honda retire; Ford sports car developments; Failure of four-wheel-drive cars; Legislation for wings; Stewart's championship in Tyrrell Matra-Ford; Ford engine domination.

10 1970: AMBITION FULFILLED 170

Year's GP races reviewed; Emergence of March Engineering; V12 BRM victory; Growing strength of 12-cylinder cars; Surtees builds Ford-powered GP car; Ferrari breakthrough; Rindt's posthumous championship for Lotus and Ford; 34 GPs and three championships for DFV engine; Duckworth goes back to his drawing board.

11 IMPROVING THE BREED 198

Four years of Ford DFV developments; Timing gear weakness; Early lubrication problems; Other first-season faults; Modified oil and water separation systems; Redesigned camshafts; Four-into-one exhaust system; Valve gear solutions; 1970 crankshaft failures; Piston modifications; Importance of trial and error; Ferrari design and development potential; Cost versus component life; Development continues.

APPENDIX 1: INDEX OF FORD DFV AND DFW ENGINES 210

APPENDIX 2: WORLD CHAMPIONSHIP SCORE-SHEET, 1966-1970 215

Acknowledgments

IT TAKES an author – or in this case two authors – to write a manuscript, but a book cannot be produced without the generous co-operation of many other people. It is in the role of publisher of Such Sweet Thunder, therefore, rather than as the book's co-author, that I record my appreciation and thanks to so many people who have collaborated in its production.

Inevitably a large proportion of them are either executives of Cosworth Engineering or the Ford Motor Company or are people who at some time during the past four years have made such good use of the DFV engine. Some are mentioned on the following pages because they are an integral part of the story, but others remain anonymous, which does not mean that their co-operation has been any the less welcomed and appreciated. I trust they will forgive me if I refrain from mentioning them all individually.

However, I must single out for special mention a few people who have contributed so much to the physical production of this book, notably of course my co-author David Phipps, whose long conversations with Keith Duckworth have provided so much of the interesting material contained in Chapters 4 and 11. David's substantial library of photographic negatives was also an invaluable supplement to Ford's own photographic files when the time came to select the illustrations for the book.

A special word of thanks, too, to Ray Hawkey, whose artistry and feeling for the subject has resulted in such a fine jacket design, also to the ever-helpful Don Gilburn who did so much to translate Ray's work into the finished product. Finally, to Mrs. June Scrace, who worked so industriously to convert 80,000 indifferently typed and scribbled words into immaculate typescript in order to meet an 'impossible' deadline and so allow me and my colleagues to remain on speaking terms with our printers.

J. B.

Introduction

THE 1967 DUTCH GRAND PRIX was drawing to a close, and only one driver seemed to matter . . . Jim Clark in his Lotus 49. Three days earlier he had not even seen his car, nor heard the sound of its brand new Ford V8 engine. Yet now he was guiding them to victory, more than 20 seconds clear of his closest challenger.

To cover the Zandvoort circuit 90 times means racing for two-and-a-quarter hours . . . a long time for sustained on-the-limit driving, and even longer when the odds are stacked against you, as they always must be when you are using a completely new car with a completely new engine. This is why the tension had increased, lap by lap, as Clark had rushed by every 90 seconds with almost clockwork regularity. It was just too good to be true . . . surely something had to break?

Suddenly, with 88 laps completed and only two more to go, Clark's engine lost its hitherto crisp exhaust note. Groans of disappointment in the grandstands and from the crowds perched high on the sand dunes lining the circuit could be heard even above the roar of the cars as the race commentator shouted emotionally in a torrent of jumbled-up Dutch, English and French, that the Lotus was in trouble. But all was well . . . Jimmy was merely easing up to conserve his fuel, knowing that his lead was safe. Three minutes later, with one hand held high, he acknowledged the thunderous applause which echoed across the track as he received the chequered flag.

Once again Clark had performed magnificently, but so had his 400 horsepower Ford DFV engine, a compact but efficient-looking power unit which had been conceived through the union of one of the world's largest car manufacturers and a brilliant young designer called Keith Duckworth.

Duckworth and his Cosworth Engineering colleagues in Northampton had been commissioned to design a race-winning grand prix engine and to manufacture five examples for supply to Team Lotus. The £100,000 arrangement constituted the Ford Motor Company's timely contribution to the future well-being of grand prix racing, the very existence of which had been threatened by an unexpected shortage of competitive power units.

Grand prix engines inevitably are very highly stressed pieces of machinery, and as such are inclined towards temperament and fragility, especially during the early stages of their development. It is a considerable accomplishment, therefore, when one of them survives its first race, and quite remarkable if it actually wins it. But the Ford DFV was to prove a most remarkable engine.

In its first season of racing it was to take Jim Clark and his Lotus to three more victories in world championship events, and to power either Clark or his team-mate Graham Hill to the fastest practice lap time in all nine grands prix in which it appeared. But this was to be only the start of the illustrious story of the DFV, the initials of which could well have stood for Duckworth For Victory rather than merely Double Four Valve (the Formula 1 engine being in effect a doubled-up version of Duckworth's FVA Formula 2 power unit).

In 1968 the engine was put into series production and made available to other car manufacturers and teams, although Graham Hill and Lotus were to win the drivers' and constructors' world championships at the end of a season marred by the tragic death of Jim Clark in a relatively unimportant Formula 2 race. That year Ford power was used to win 11 of the 12 world championship races, Hill's strongest challenger for the title being his former BRM team-mate Jackie Stewart, who was now driving a Ford-engined Matra.

Stewart's burning ambition to win the world championship was to be fulfilled the following year, when all 11 championship events were won by Ford-engined cars, Stewart's Matra being the first across the finishing line in six of them. By this time the DFV engine was dominating grand prix racing so completely that it was associated with no fewer than 261 of the 275 world championship points scored by drivers and all but 14 of the 211 collected by constructors during the 1969 season.

To be involved with both world titles two years in succession was a magnificent achievement, especially as the Ford V8 was facing opposition from 12-cylinder engines which theoretically should have been capable of beating it. But the DFV success story was far from over, and 1970 was to see the completion of a world championship double hat-trick for the Ford engine, a brilliant achievement which was to be marred by the cruel death of Gold Leaf Team Lotus' number one driver and new title-holder Jochen Rindt. The Austrian's five victories with his Lotus-Ford prior to his accident at Monza were sufficient to give him the world championship – his life's ambition – despite the very strong challenge to the Ford-powered cars which was offered by Ferrari during the second half of the year.

By the end of its fourth season of racing Keith Duckworth's engine had taken part in 45 world championship events and had won 34 of them. It had been fitted to at least 10 different makes of grand prix car, had been seen in sports car endurance racing, and as a smaller-displacement DFW version had won honours in Tasman races in New Zealand and Australia.

Cosworth Engineering, who had set out to produce five engines for one team (in fact they produced seven in the first year) went on to build a further 70 during the next three years, and then laid down an additional batch for the 1971 season incorporating the first major design modifications aimed specifically at increasing power. It had taken the opposition between three and four years to catch up, and now that they had managed to close the gap Keith Duckworth was mounting his own counter-attack to continue the power battle which had been such a fascinating feature of the 3 litre grand prix formula.

This is the story of the Ford DFV engine, a brilliantly executed piece of machinery measuring less than 22 inches long and 27 inches wide, weighing some 360 pounds and selling for £7,500; an engine which gave grand prix racing an urgently needed transfusion in 1967 then went on to provide its lifeblood for several more years. It is also the story of the people who conceived it, who built, sponsored, raced and won with it, and who on occasions even broke it. It is a story without an ending, for the Sweet Thunder of the DFV's exhausts will echo around the race circuits of the world for many years to come, and even eventually when they are silenced there will still be a special place in motor racing's Hall of Fame for this remarkable piece of engineering and the people associated with it.

1

The lesson of 1961

ONE OF THE MOST frequently heard phrases in a motor racing paddock is 'You should have seen how old so-and-so went past me down the straight'. It is the racing driver's standard remark – perhaps excuse is a more apt definition – after failing to set a fast lap time or win a race. It infers that his car was underpowered, and that all the time which he had gained through the corners by his brilliant driving had been sacrificed down the straights for the want of more horsepower.

But extra horsepower is not the only reason why one car may pass another on a straight piece of track. The faster car may have gained its extra speed because its driver, either through greater skill or more suitable gearing, was able to leave the previous corner that much faster. And if the passing manoeuvre took place towards the end of the straight it was probably aided by the driver of the second car tucking into the slipstream of the car ahead and getting a 'tow' in the thinner air before moving out again into the air stream to pass.

But if any driver of a British Formula 1 car uttered those well-used words during the 1961 grand prix season, few people chose to question their authenticity, or to doubt that 'old so-and-so' went past because he had more horsepower . . . quite a lot more. For Britain was once again in the grand prix doldrums, a position with which she had been familiar throughout much of the history of international motor racing, but from which she had finally emerged during the late fifties.

A tremendous effort by a number of relatively small companies and a much larger number of dedicated motor racing enthusiasts, who also happened to be very fine engineers, mechanics or drivers, had combined to put Britain on top in the racing world for the very first time. The breakthrough had occurred in 1968, when the Vanwall, designed and built near London by Tony Vandervell's bearing company, had emerged as the world champion car, and thanks to the concerted efforts of the Cooper Car Company, Coventry Climax Engines Limited and Jack Brabham, the country remained very firmly on top throughout 1959 and 1960. Yet in 1961 Britain was back in the grand prix wilderness, powerless to prevent Italy's Ferrari team run away with race after race with almost contemptuous ease. And it all happened through shortsightedness.

It has been the policy of the international governing body of motor sport, the *Federation Internationale de l'Automobile* (*FIA*), through its sporting sub-committee, the *Commission Sportive Internationale* (*CSI*), to reframe the regulations governing the various categories of motor racing every few years. The prime intention of this policy is to prevent design stagnation, and to stimulate and encourage technical advancement by periodically providing designers and engineers (not to mention racing drivers) with a new type of challenge.

But from time to time regulation changes have been imposed for political reasons, and there is no doubt that this applied when, towards the end of 1958, an announcement was made regarding the new regulations which were to be introduced for the grand prix Formula 1 as from the beginning of 1961.

As had been the case for many years, the basis of the regulations was to be a restriction on engine capacity. Since the beginning of 1954 the limit for unsupercharged engines had been 2½ litres, but starting in 1961 this limit was to be reduced to 1½ litres, and furthermore there was to be a minimum-weight restriction on Formula 1 cars, aimed at preventing dangerously lightweight construction.

The decision substantially to 'detune' grand prix cars to a level equal to the then current Formula 2 machines had been prompted by the CSI's concern about the climate of public, and in some instances governmental opinion regarding the many accidents which had been marring motor racing during the previous few seasons. The great Le Mans tragedy was still only three years past, and memories of its enormity were still painfully vivid. The fact that the major contributions to the Le Mans accident situation had been the narrowness of the track and the wide differential in the speeds of different cars at that point on the circuit, rather than sheer engine size or performance was irrelevant in the eyes of the CSI, since in their view governmental and public opinion was swayed by emotion rather than by technical analyses. Also, rightly or wrongly, the severest critics of motor racing were always inclined to relate accidents to speed, and speed to engine size.

Equally, the CSI were mindful of the fact that within the four months preceding their decision regarding the 1961 formula, three grand prix drivers, Luigi Musso from Italy and Peter Collins and Stuart Lewis-Evans from England, had lost their lives after their Formula 1 cars had gone out of control during a world championship race. In the circumstances they were in no mood at that time to consider any change which would increase the performance of grand prix cars.

Nevertheless, the decision brought an outcry of protest from the British teams, various spokesmen of which put forward a number of valid technical reasons to support their view that the 1½ litre formula as proposed could, under certain circumstances, actually increase rather than lessen the dangers of grand prix racing, but then they virtually killed any chance of cool debate by themselves indulging in emotion and announcing that they would boycott the new formula.

This news must have been music in the ears of Enzo Ferrari, who, having been removed from the pinnacle of motor racing by Tony Vandervell, and subsequently kept down by John Cooper, Leonard Lee and Jack Brabham, was understandably anxious to regain his own and his company's lost stature, and saw in a change of Formula 1 regulations – any change – the opportunity to do so.

Ferrari set his engineers to work at once to design a new car, together with two versions of a 1½ litre V6 engine with which to power it, one with cylinder banks angled at 65 degrees, the other with the cylinders at 120 degrees. The British companies, meanwhile, absorbed themselves in the more immediate problems of retaining their leadership of grand prix racing during the two remaining years under the 2½ litre formula, in which they were eminently successful, despite the fact that Tony Vandervell had

ERRATUM

Page 12, fourth line

For '1968' read '1958'.

withdrawn his Vanwall team from the scene following his 1958 success. Only when it was far too late did the British racing teams accept the fact that the 1½ litre formula would be operated with or without their support, and turn to the task of designing and building the necessary cars and engines.

It is by no means uncommon for a completely new racing car chassis to be designed and built within the space of a few months, but it is an altogether more difficult and lengthy business to design, construct, test, develop and put into production a completely new racing engine.

With the withdrawal of Vanwall, this task was to fall on just two British companies, the Owen Organization's research department and BRM facility at Bourne, in Lincolnshire, and Coventry Climax, who had been engine suppliers to both the Cooper and Lotus teams under the 2½ litre formula. Both companies laid down designs for V8 Formula 1 engines but neither of them stood any chance of producing them in time for the start of the new formula, and Britain would have had few if any cars on the starting grids for most of the 1961 races had it not been for the fact that Coventry Climax already had a 1½ litre four-cylinder engine on the shelf which they had been supplying to customers for Formula 2 or small-capacity sports cars. Fortunately, they were able to rush through a quick development programme on this relatively unsophisticated engine to give it a little more power, and then build sufficient of them to equip all the British teams, including BRM, until such time as their own and BRM's new V8 engines were ready. The four-cylinder Coventry Climax engine was to keep Britain's grand prix racing wheels turning throughout the 1961 season, though considerably more slowly than the Ferraris' in most races. It was a frustrating period, and a measure of the magnitude of Britain's mistake in ignoring the new formula proposals instead of making the best of them. It was also an indication of the extent to which British racing teams, unlike their Continental rivals, were dependent on an outside supplier for their engines, a situation which was to have a certain parallel, but for very different reasons, when the next change in the grand prix racing formula took place five years later, and the Ford Motor Company and Cosworth Engineering came to the rescue.

Despite all the forebodings, the 1½ litre formula was to prove very successful, and within two years most of the cars were lapping the grand prix circuits faster than the best 2½ litre cars had managed in 1960. Admittedly, it was no longer quite as difficult for a driver of average ability (by grand prix standards) to extract a good performance from his car as it had been in the 2½ litre days (this had been one of the main points of criticism of the new regulations), but despite the reduction in horsepower and the continuing improvement in chassis design and roadholding, drivers of outstanding calibre such as Stirling Moss and Jim Clark still had little difficulty in demonstrating their superior talents and taking full advantage of them.

It was the genius of Moss, in his last full season as a grand prix driver, which prevented 1961 from being a complete debacle for Britain, for despite giving away between 20 and 30 horsepower to the Ferraris of Phil Hill and Wolfgang von Trips, he won both the Monaco and German Grands Prix in Rob Walker's Lotus-Climax, after two of the most outstanding performances of his illustrious racing career. Then, when Ferrari decided to stay at home having already won the world championship, Innes Ireland was able to add a third British victory in 1961 when he gave Team Lotus their first-ever grand prix success at Watkins Glen, in the United States. Nevertheless, Ferrari had won five world championship races in seven attempts.

It was perhaps fortunate for Britain that Ferrari was the only rival team to present a fully developed and reliable new engine in 1961, and that Porsche, who were in considerable difficulties with their new flat-eight engine, were forced to fall back on their former Formula 2 four-cylinder power unit, the performance of which was roughly on a par with the 150 b.h.p. Coventry Climax. Had the new Porsche engine escaped its teething troubles, 1961 would have been an even bleaker grand prix year for Britain.

Although the first Coventry Climax V8 engine was seen in action in Jack Brabham's Cooper at the German Grand Prix, and reappeared for the two remaining world championship races, it was never to finish a race that year, while the BRM V8, which Graham Hill first tried in practice for the Italian Grand Prix, was not used in a race during 1961, nor did Moss start a race with the Climax V8 which had been made available for him late in the season. However, both engines benefited from intensive development programmes during the winter of 1961–1962, and thanks to some dedicated work at Bourne and Coventry Britain was able to regain much of her lost prestige when the world championship battle was rejoined in the spring of 1962, a battle in which Stirling Moss, alas, was to take no part as his very serious accident at Goodwood in the UDT-Laystall Lotus-Climax brought his racing career to a close and so nearly cost him his life.

Moss's absence caused the 1962 championship battle to develop into a straight fight between Jim Clark, leading Team Lotus, and Graham Hill, the number-one driver of the BRM team, with their main opposition coming from Bruce McLaren, who had been elevated to the position of Cooper team-leader following Jack Brabham's decision to build and race his own cars.

The championship was only decided in the last race of the year, in South Africa, where Clark, having started favourite and streaked into a big lead, bequeathed both the race and the title to Hill when a bolt vibrated loose from his engine's distributor drive housing and let out all the oil. There were nine world championship races that year, of which BRM won four, Lotus-Climax three, Cooper-Climax one, and Porsche one, while Ferrari, so all-powerful the previous year, failed to score a single victory; British engine power had been a late starter but at least had made a triumphant come-back.

1963 was the year of the big revenge for Jim Clark and the Lotus-Climax, for together they amassed no less than seven victories out of the ten world championship races, two of the remainder going to Graham Hill and BRM, and one to Ferrari, who now had John Surtees as their new team leader.

Surtees was to prove a great tonic to the Ferrari team, who had become very depressed by their poor showing during the year following Phil Hill's championship success, and in 1964 'Big John' was to give them two more victories in an unusually open season at the end of which he emerged a surprised world champion, just one point ahead of Graham Hill and eight ahead of Clark, whose three race victories in the first half of the season were followed by a barren four races during which he failed to add a single world championship point. Once again there were ten qualifying races, of which three went to Ferrari, three to Lotus-Climax, two to BRM, and two to Brabham-Climax, Ferrari emerging as narrow winners of the constructors' championship.

They say that nothing succeeds like success, but this did not seem to apply to the Ferrari team in the Sixties, for just as Phil Hill's world championship had been followed by a year in the GP wilderness, three years later John Surtees' title was to be followed by a season of greatly diminished fortune both for himself and for his team. 1965, like 1963, was Jim Clark's year, although this time he was able to put his Lotus-Climax first past the post only six times out of ten. Hill's BRM was the winner of two other races, Jackie Stewart's BRM scored one victory, and right at the end of the season, in the final race under 1½ litre regulations, Richie Ginther pulled off a surprise and popular victory in Mexico with the raucous V12 Honda, the Japanese team having put in a tentative appearance at the Italian Grand Prix in 1964 before making their first full-scale effort in GP racing with a two-car team in 1965.

The Honda appearance had been welcomed universally in grand prix circles, because although the 1½ litre formula had proved most successful after its shaky beginning, it needed the added stimulus of more international competition. Porsche had withdrawn from the scene at the end of 1962, and although the gap in the ranks had been filled temporarily by a new Italian team, ATS (which had been formed by a set of disgruntled ex-Ferrari employees), their car proved hopelessly uncompetitive and soon disappeared again. Another car destined to make only a brief appearance was the Scirocco, a conventional but rather heavy design powered by the BRM V8 engine and financed by a consortium of Anglo-American backers, while the British Racing Partnership, which hitherto had been racing Lotus cars, decided to build their own BRP monocoques, which were also fitted with the BRM engine.

Ferrari had managed to counter-attack the breakthough by the British V8-engined cars by replacing their V6 power units with a V8 for the 1964 season, and whereas the previous year the Ferrari engines had been cracking in trying to match the British V8s, which were now giving around 200 b.h.p., in 1964 it was the Climax which ran through a period of fragility. But the setback was only temporary, and although Ferrari produced a flat-12 to supplement their V8 in 1965 they found themselves outclassed once again by the final version of the Climax V8, with its four valves per cylinder, and a power output in excess of 220 b.h.p.

However, Ferrari did not seem to be unduly perturbed, because the 1½ litre formula was drawing to a close, and he was more concerned that his team should have the best 3 litre power unit (and car in which to instal it) when the 'new look' grand prix field lined up for its first race at Monaco in May 1966. The Ferrari team had been well prepared for the start of the 1½ litre formula five years earlier, and were determined to be equally ready this time.

The change in formula had been notified over two years earlier, and on this occasion there had been few if any dissenters. Perhaps the five years of 1½ litre racing had had a useful therapeutic effect because it had proved a relatively safe formula, and as a result the climate of public opinion seemed prepared to accept that a return to more powerful grand prix cars was now not only acceptable but necessary if Formula 1 was to retain its status as the premier form of motor racing. After all, were we not in the age of the 7 litre sports coupes, racing down the Mulsanne Straight at Le Mans at speeds of well over 200 m.p.h.?

The new limit obviously suited Ferrari because they had a wealth of experience of building 3 litre engines, while Honda clearly were intent on supporting the new formula, otherwise they would not have ventured into grand prix racing in the first place. They had built their 1½ litre cars and engines in order to obtain practical experience of Formula 1 racing and to examine the strength of the opposition from close quarters, and when they had first unzipped their kit bags in Europe they seemed to reveal as many cameras as spanners!

Although there was no sign of any imminent return to grand prix racing by either Mercedes-Benz or Porsche it seemed likely that the doubling of the engine displacement limit would increase rather than decrease the possibility of such an event, while in Italy both Alfa Romeo and Maserati had suitable engines either in existence or on the drawing board and so might well be persuaded to become part of the grand prix scene once again.

It seemed likely, therefore, that the British racing teams would be meeting some very tough new opposition under the new formula, which in one respect was welcome news because persistent domination of Formula 1 racing by any one country is not in the best interests of the sport or even, in the long term, for the country which is on top. Grand prix racing, by its very nature, thrives on competition between nations as well as between individual drivers, cars and engines.

Clearly, whatever the strength of the opposition from overseas, British teams and drivers would once again be looking to BRM and Coventry Climax for their power units, and during 1965 BRM announced that they had investigated two possible layouts for their proposed 3 litre grand prix engine. One was to add four cylinders on to the end of their 2 litre Tasman engine (this was an enlarged version of their Formula 1 V8 for use in the Tasman series of races in New Zealand and Australia), thereby converting it into a 3 litre V12. The alternative idea was to mount two 1½ litre eight-cylinder engines on top of each other and geared to a central crankshaft. In this case the cylinders were to be horizontally opposed in banks of four, instead of angled in a Vee, so as to form an H16 layout. After careful consideration of both types they had chosen the latter arrangement, they said, because it would provide a more compact power unit and would enable them to adopt similar cylinder head and valve gear components and layouts to those used in their 1½ litre V8s, about which they had amassed a great deal of data. On paper it seemed to be an ambitious project, although whether or not the engine could be made to work remained to be seen.

Although the company had been taken over by Jaguar Cars Limited in 1963, Coventry Climax, it was assumed, would continue to supply engines to the Lotus, Cooper and Brabham teams. For many years the company's main business had been the design and manufacture of industrial and marine engines, but it had slid very smoothly into the very specialized world of racing engines from that day in the mid-Fifties when someone had discovered that a Coventry-Climax fire pump engine, suitably modified, was a most effective way of propelling an 1,100 c.c. sports car.

Since then their impact on racing had been immense, at first by powering lightweight sports cars to a bewildering number of successes, then in Formula 2, where again they became the power suppliers of most of the British cars, and eventually in Formula 1, where they had given Jack Brabham and Cooper the horsepower with which to win their world titles in 1959 and 1960.

During the five years of the 1½ litre grand prix formula which followed they had encountered early problems but had quickly mastered them to enjoy long periods of success during which they had helped to win two more world titles for both drivers and constructors. They had also built for themselves an invaluable fund of knowledge and experience of racing engine design and development to make them a considerable authority in this field. It was assumed that they would remain so under the revised regulations to be enforced for the 1966 season.

Inevitably there were rumours and counter-rumours about the form which the new Climax GP engine would take. The 'simple V8' theory was prompted by the knowledge that the company had had an exciting 2½ litre V8 ready for use in 1960, which had been placed under a dust sheet when the GP formula had changed to 1½ litres. It would be a comparatively simple matter to enlarge this engine to a full 3 litres.

Another rumour suggested a 16-cylinder engine (perhaps with the V8 being used as a temporary stop-gap until the more complicated engine was fully developed). The origin of this particular piece of crystal ball-gazing was presumably the awareness in certain quarters of the existence of a Coventry Climax flat-16 1½ litre engine, of which four examples were to be built and loaned to teams for the 1965 season (in fact this project was dropped because the latest 32-valve V8 was to prove more than adequate for the job). Such an engine in 3 litre form was an exciting prospect if you tried not to think too hard about its bulk and weight.

However, there was to be neither a V8 nor a flat-16 from Coventry Climax Engines Limited. Early in 1965 the company shattered the racing world by announcing that they would be withdrawing from all forms of racing at the end of the year, apart from making available spare parts and a certain amount of service for existing engines. They had signified their intention to pull out of racing on previous occasions but each time had decided to carry on, despite the very considerable cost to the company. However, this time there was to be no turning back. Racing, they claimed, had cost them something between £500,000 and £1,000,000, and this sort of financial burden could be carried no longer. But it wasn't just the money; motor racing had been too demanding on the time of their technical staff, and other research and development projects had been suffering. One could sympathize with their predicament, and perhaps could agree that their decision was in the company's best interests. But this did not lessen the body-blow which was dealt to Britain's future grand prix aspirations on that dark day in February 1965. Lotus, Cooper and Brabham, all of whom were planning a full-scale assault for this exciting new period of grand prix racing, and each bursting to prove that they could build the best chassis to cope with 400 or more horsepower, suddenly found themselves in the melancholy position of having all the chassis design ideas in the world, but not a horsepower between them. The Coventry Climax bombshell had arrived when the first 3 litre Formula 1 race was less than 11 months away; and ironically the new Formula was being billed as 'the Return of Power!'

2

Any other business

WALTER HAYES' sole contact in the world of motoring and motorsport prior to the day he joined the Ford Motor Company was Colin Chapman, and the only reason he knew Colin was that as editor of the *Sunday Dispatch* he had employed him as a free-lance motoring correspondent.

Walter's appointment to the post of Director of Public Affairs, Ford of Britain, was a direct result of the close friendship which existed between Ford's dynamic Chairman, Sir Patrick Hennessey, and the equally dynamic Lord Beaverbrook. Sir Patrick knew that Beaverbrook was an unusually shrewd judge of men, and when, during the course of a major reorganization of the Ford management in 1961, he needed someone to fill this last remaining top position, he asked the newspaper magnate if he knew of anyone capable of handling the job. Unhesitatingly, Beaverbrook recommended Hayes.

When Walter joined the company he found that life was full of surprises, particularly in the area of personal responsibilities. He discovered, for example, that one man was doubling the duties of competitions manager and publications manager, which seemed to him to be a most effective way of ensuring that neither job received full-time attention.

Hayes lost little time in applying some remedial treatment, and his cure for this particular ailment was to tell Sir Patrick that he felt that Competitions should cease to be the responsibility of the Public Affairs department. It is possible that there was also a slight ulterior motive behind this move, because if there was one subject on which Walter Hayes was less well informed than he was about cars it was motorsport.

But Hayes had many other qualities which more than offset his limited knowledge of certain subjects. Notably, he emerged very quickly as an outstanding 'ideas' man, with a flair for breaking through cobweb-covered conventions into new and imaginative areas of public relations activity. An articulate speaker, he was able to sell his ideas with enthusiasm and conviction. His enthusiasm became contagious both within and outside the company, and as you watched him expound on his Idea of the Moment, his large eyes burning brightly and his hands rubbing themselves together faster and faster as if they were demonstrating an exciting new line in toilet soap, it was difficult not to be moved by the latest Hayes crusade, which perhaps was fortunate because almost invariably it turned out to be a winner.

Inevitably Sir Patrick became well pleased with the progress made in the field of Public Affairs during the year following Hayes' appointment, but he was less happy that Walter had talked him into relieving him of the responsibility for the competitions department.

Accordingly, one day early in 1963 Sir Patrick sent for Hayes and informed him that Competitions was to be passed back into the orbit of Public Affairs. Walter accepted the news with some misgivings, for he was still unconvinced that it was a good idea, but this time Sir Patrick was adamant. Probably neither man realized it at the time, but that moment was one of infinite importance to the Ford Motor Company, the ramifications of which during the decade which followed were to be seen and felt not only in Britain but throughout the world, or at least throughout those parts of it where cars and drivers were brought together in the name of motorsport.

Good timing is a tremendous asset in the successful conduct of business, and there is little doubt that Walter Hayes could not have timed his 'takeover' of Competitions more precisely, even if he had planned and engineered the whole operation himself. He was also fortunate in the knowledge that 'performance' was deeply embedded in the essential character of the Company. The name of Henry Ford I first made the headlines when he won a motor race and one thousand dollars with a car of his own design. Henry Ford

II, his brilliant grandson, had personally led Ford of America back into motor racing in the Sixties with a campaign which resulted in domination at Indianapolis and triumph at Le Mans.

Inside Ford of Britain there were men like the Chief Engine Designer Alan Worters, Foundry Manager Tru Hayford and top Manufacturing Directors such as Andy Taylor and Sam Rees who would always find time to advise motor racing men from inside and outside the Company and make 'special bits', while Sir Leonard Crossland, now Chairman of Ford of Britain, has always had a private stable of fast cars. These men knew that one of the reasons that Ford cars and engines were good on the race track was that they were 'made of good stuff'. Hayes therefore had the right raw materials and the right kind of allies.

The competition department became his to control at the very moment when the Cortina was emerging as the most successful model Ford of Britain had ever manufactured.

Already tuning establishments up and down the country, and overseas, were working on the Cortina and its engine in preparation for various forms of competitions activity. The original Cortina had had a 1,198 cc engine, but a 1,498 cc version with a new cylinder block to carry a five-bearing crankshaft had been added to the range, and now a higher-powered version of this larger engine, utilizing a Weber compound-choke carburettor and improved breathing, had been developed to power the first Cortina GT. This was the car which was to launch Ford of Britain into a race and rally programme which proved so successful that its momentum virtually committed the company to continuing and expanding involvement in motorsport, a policy which received further stimulus at the end of 1963 by the introduction of the Cortina Lotus with its twin-overhead-camshaft engine. In a very little time the department which Walter Hayes had reinherited as a 'promising sideline' became one of the most valuable weapons in the Public Affairs armoury. The company had conceived an extremely successful 'bread and butter' car which, with the aid of a carefully woven montage of technical expertise, driving skill and imaginative promotion and exposure, was soon to become one of Europe's most successful and versatile competition cars, and with an extrovert at the helm Ford's Public Affairs people knew exactly how to extract the maximum mileage out of the success story.

The Cortina, more than any other car, had given Ford a competition image, an image which not only was welcomed in motorsport circles, but was to prove a magnet wherever Ford cars were sold, drawing into dealers' showrooms in increasing numbers car buyers who, though they had never owned or even aspired to possess a competitions licence, wished to bask in the reflected glory of driving a car which was at least visually identifiable with Cortinas raced at Brands Hatch or Silverstone or rallied from Oslo to Monte Carlo.

Walter Hayes had been quick to appreciate the value of competitions successes to Ford's corporate prosperity, but at the same time he was anxious that the company should not be thought of as one which took a lot out of motorsport but failed to put much back in. Remembering that you were never more vulnerable than when you were at the top, Hayes was determined that, in so far as he was able to control these things, Ford's involvement in any particular branch of the sport, if successful, would not be maintained to the point where all effective opposition was frightened away, and the sport allowed to suffer through lack of non-Ford competition.

Unfortunately, this laudable aim is not always easy to achieve, but Ford can scarcely be blamed for its domination of motorsport in the Sixties, for no company has the power to control the ability or otherwise of a rival organization to match its performance on the race track or the rally route. It can only do its own job to the best of its ability and hope that at least one other company will do likewise and prove almost – but not quite! – as successful. When all effective opposition fades away, or indeed fails to materialize at all, a company cannot suddenly abandon a motorsports programme just because it is doing all the winning; some return has to be seen for the considerable investment in design ingenuity, technical expertise and perhaps even fundamental research which has been put into it. The rewards to be won by a car manufacturer from active participation in motorsport are not measured just in terms of the amount of cash and the number of garlands won, or even by the number of extra sales in the market place attributed to these successes (even if they can be calculated accurately, which is debatable), but also by the technical lessons which are learnt from racing or rallying, which can be even greater during moments of failure than they may be at times of success.

Nevertheless, it has always been part of Ford's philosophy of motorsport involvement that they ring the changes as regularly as possible so as to avoid either excessive domination or development stagnation, or both. Also, as far as is practicable, they will conduct their motorsport activities in close collaboration with specialist independent teams or organizations which already have a proven record of success in the sport, whether it be as active competitors, competition car builders, engine and chassis tuners, or any other activity involving specialized talent, skill and experience. But in all these associations they maintain a two-way 'hot line' of communication between the factory-assisted team or organization and Ford's own competitions department in Britain, Germany or any other country where the company may from time to time establish their own competitions facility.

Many examples exist of close and successful associations which were formed during the Sixties between Ford and independent motorsport organizations, such as the operation of saloon car racing programmes by the Alan Mann Racing Organisation and by Team Lotus in pursuit of the European and British saloon car championships respectively, which left the company's own competitions department free to put their own main effort into international rallying. In an entirely different field there is the example of Ford laying the foundations for a long and highly successful venture in endurance racing before passing it over to an independent operator. This was the policy in respect of the Ford GT40, much of the design, development and manufacture of which was carried out at Ford's Advanced Vehicles facility at Slough under the direction of John Wyer, until Wyer was given the opportunity to take over the premises and continue to manufacture, service and race the car under the banner of his own J. W. Automotive Engineering company. This was Ford's way of ensuring that owners and prospective owners of GT40s would continue to get all the technical backing and maintenance and spares facilities which are so essential in racing while Ford's own personnel were left free to concentrate on their next project.

Inevitably, much of Ford's European-based competitions programme has been confined to events within the Continent, but there is also a history of successful participation in selected races and rallies in other parts of the world, notably in those areas where a large market exists for Ford's European-built cars or is being developed. There was, for instance, the occasion in 1963 when both the Willment and Alan Andrews racing teams, under the management of Jeff Uren and Alan Mann respectively, were encouraged to take two of their Cortina GTs across the Atlantic to Marlboro, Maryland for a particularly well-publicized 12-hours saloon car race, which hitherto had been the happy hunting ground of rival manufacturers; they responded by finishing first and second overall. Across the equator, the African market for Ford cars was greatly stimulated by the company's various successes in the East African Safari, notably in 1964, when Cortina GTs finished first, third and fifth overall, and here again is an example of Ford's European-based competitions departments working closely with their local teams, in some cases by mounting a tandem operation and in others by allowing the locals to act as the official representatives of the European factories.

More recently, Ford's enthusiastic participation in the new breed of across-the-world rallies has been predicated on the knowledge that these events pass through many countries which are already profitable export markets and through others which have important future market potential. The company obtained a great deal of favourable publicity from their 'near miss' in the 1968 London-to-Sydney Marathon by the Roger Clark/Ove Andersson Cortina Lotus, which was converted into extra sales in many countries, but this was nothing compared to the sales boost which followed the sweeping victory of the Escorts, headed by the Hannu Mikkola/Gunnar Palm car, in the 1970 World Cup Rally from London to Mexico, a success which resulted in the birth of an additional model in the Escort passenger car range.

Although the Cortina was Ford's major headline-maker during the mid-Sixties, and the car which brought some badly needed new life into various areas of motorsport, its contribution to the sport was not to be measured in this alone. Ford's desire to contribute something tangible to the future well-being of the sport was to be fulfilled following the announcement by the FIA that as from 1964 the 1,100 c.c. Formula Junior class of single seater racing would be replaced by two classes, Formula 3 and Formula 2. Although both were to have a capacity limit of 1,000 c.c., there was one important difference; whereas a Formula 3 engine, like its

Formula Junior predecessor, was to be virtually a race-tuned version of a production car power unit, the Formula 2 regulations permitted the construction of special racing engines, subject to certain limitations on design and with the proviso that the cylinder block should come from a production car.

Keith Duckworth, whose Cosworth Engineering company's rapid growth had been built on his very successful adaptation of the Ford Anglia engine for Formula Junior racing, was convinced that the same power unit should also form the basis of his new Formula 3 power unit. But as the regulations governing Formula 2 offered him a great deal more scope, notably in the field of cylinder head design, he made some careful studies of all available cylinder blocks before deciding which was the most suitable as the basis of his projected Formula 2 engine, which he intended would have valves operated from a single overhead camshaft, and for which he was anticipating a power output in the region of 120 b.h.p. He chose the Cortina block, and after Walter Hayes had listened to Duckworth's plans for the new formula he offered him Ford backing to the extent of £17,500 as a contribution towards the cost of designing and developing his SCA engine, in return for which Duckworth undertook to inscribe both the Cosworth and the Ford names on the valve cover. This was the first occasion on which Ford had made a financial grant to an independent engine designer in Europe for a project other than one involving the company's own research and development activities.

Walter Hayes by this time was already a great admirer of Duckworth, not only for his prowess as an engineer, but also for his incisive mind, his forthright manner and his dedication to his work. Although the two men were so different in many ways they probably found in each other a kindred spirit, for both had a deep dislike of doing things according to the book, and they shared each other's ability to analyse a situation in simple terms with a clarity of thought free from the prejudices of convention. They may not have known much about the technicalities of each other's profession, but somehow they seemed to talk the same language and soon developed a deep respect for each other, even during their occasional moments of bewilderment or even disagreement.

Duckworth's SCA engine did much to consolidate the growing reputation of Cosworth Engineering, even though it was to meet very tough opposition from the Honda power unit duriug 1966, the last of the three seasons during which Formula 2 was confined to a 1,000 c.c. capacity limit. Nevertheless, there were certain people who appeared to read too much into the eventual eclipse of the SCA by the very powerful and very expensive Hondas used by the Repco Brabham team, failing to realize that it was senseless for Duckworth to spend a lot of time, effort and money in order to match the Honda's performance when the 1,000 c.c. engines would be obsolete within a few months.

These were the people – and they included some quite exalted names – who, when it was revealed that Ford was about to back Duckworth and his colleagues on an infinitely more ambitious project of engine design, suggested that the company would live to regret their decision. Their inference was that although Duckworth was undoubtedly a highly proficient modifier of engines he had still to prove himself as a racing engine designer. If Duckworth himself heard any of these murmerings the only effect they had on him was to make him even more determined to do a good job, while those most closely involved with the project at Ford were also unperturbed; they believed firmly in the concept of backing people and following their faith in individuals, and they were convinced that Duckworth would not let them down. How right they were!

Many stories have been written or told concerning the moment at which the idea of a Ford grand prix engine was born, the majority of them being contradictory and some of them downright false. The truth is that, like so many of Ford's competition activities, the idea just evolved through a careful appraisal of circumstances and a subsequent awareness of a need.

In this instance the circumstances were caused by the unexpected withdrawal of Coventry Climax Engines Limited from all racing activities just a few months before the introduction of a new grand prix formula. It required little imagination to identify the need of the Brabham, Cooper and Lotus teams for alternative sources of supply, but there was no precedent for a major car manufacturer, whose prime function was the design, construction and sale of passenger cars and trucks in very large quantities, creating such a source. Ford was already very heavily committed in motorsport in Europe, and if the company had chosen not to involve itself in the very specialized field of grand prix racing it could have justified its decision very easily.

Nevertheless, here was a situation in which a major part of Britain's motor racing effort was being threatened with impotency through the lack of an effective power unit. If Ford wished to make a very substantial contribution to the welfare of motorsport, and at the same time to safeguard the immense prestige which British grand prix successes had brought to the country's motor industry as a whole during the late-fifties and early sixties, this was an area in which it could be made. It was an awareness of this, more than any other single factor, which helped to carry the idea of a Ford-sponsored grand prix engine from the embryonic stage through to fruition.

Throughout his period with Ford Walter Hayes had maintained a close and friendly association with Colin Chapman, whose ability to expound eloquently, forthrightly and creatively on any subject appertaining to cars and motorsport he had found both stimulating and educational. It was understandable, therefore, that he began to see Ford's possible involvement in grand prix racing in terms of providing a series of engines specifically for Team Lotus, the seed of this idea inevitably having been sown by Chapman himself.

Shortly before the Lotus chief moved his factory and his home to Norfolk he invited Hayes to dinner at his house at Hadley Wood, Hertfordshire, where he talked to him with considerable passion about the contribution which Britain's motorsport successes had made to British industry and in consequence to the country's balance of payments. He went on to say how essential it was, not only for Team Lotus but for all the British-based grand prix teams, to be able to maintain the momentum of international success and prestige which they had built up after their disastrous start under the 1½ litre formula; they had paid dearly for their unpreparedness in 1961, and now that they had fought back on top they were in grave danger of handing everything back to Ferrari again on a plate.

Chapman, of course, like John Cooper and Jack Brabham, had been investigating all possible sources for his next grand prix engine, but what worried him particularly was that, almost without exception, everyone seemed to be building unnecessarily heavy and complex engines, which was entirely contrary to his own concept of the ideal 3 litre Formula 1 car. He was convinced that the correct approach was to produce a minimum-weight car with a good aerodynamic form and low frontal area and to power it by a compact and relatively simple engine with a wide torque range. Such a car, he felt, would have an adequate power-to-weight ratio, and would be both easier and less tiring to drive than the heavyweight designs which certain of the existing or proposed power units would create.

Walter Hayes listened to Chapman's ideas with growing interest, and shortly afterwards he was summoned to Sir Patrick Hennessey, who told him that Chapman had approached the Society of Motor Manufacturers and Traders, of which Sir Patrick was President at that time, with a request that the SMMT should support British motor racing interests by sponsoring a national grand prix engine. When Sir Patrick asked Walter Hayes for his reaction to the idea he replied that in his view an SMMT engine had to be a disaster.

However, the Society did not ignore Chapman's request entirely, and Sir Patrick used his office to approach the British Government in the hope that they could be persuaded to 'do a de Gaulle', and provide some financial backing for a British racing effort in the interests of national prestige. The most distressing outcome of this particular piece of lobbying was not that the request was turned down but that there was a complete and seemingly impenetrable lack of understanding of the whole problem in Whitehall. Either the Government was utterly unaware of the contribution made by motorsport successes to the country's prestige and economy, or it was completely unmoved by it, and by the thought that it might cease.

Meanwhile, Walter Hayes had been discussing the idea of a Ford-backed grand prix engine with Harley Copp, a quiet-voiced smiling-faced American who had joined Ford of Britain as Vice-President in charge of Engineering, and whose burning enthusiasm for motor racing Hayes had striven to encourage. From an engineering as well as a company prestige aspect, Copp thought that the idea made a lot of sense, and he had a long chat with Chapman during which the Lotus boss again expounded his concept of a 3 litre grand prix car of the utmost compactness and simplicity. Copp was impressed.

Further discussions took place between Chapman, Copp and Hayes during which there was unanimous agreement that there was only one man for the job, provided the necessary finance was forthcoming – Keith Duckworth. Chapman had already broached the subject with Duckworth in general terms, and indeed they had already engaged in preliminary, albeit fruitless discussions with another possible sponsor within the British motor industry. He was convinced, therefore, that Duckworth would be willing to undertake the formidable job under certain conditions, one of which was that he would first make a 1.6 litre Formula 2 engine, based on the Cortina block but with twin overhead camshafts and four valves per cylinder, and convince himself that it was up to expectations in terms of performance and reliability before committing himself to doubling it up into a 3 litre Formula 1 V8. Throughout the various discussions involving Chapman, Copp, Duckworth and Hayes not once had there been any serious thought that the engine, if given the go-ahead, would be anything other than a V8. The major attraction, of course, was its compactness and relative simplicity, coupled with Duckworth's conviction that eight cylinders would be adequate for an initial target power output of 400 horsepower. There was also the convenience that the proposed FVA Formula 2 engine – which Duckworth was determined to produce in any case – would form the basis of a Formula 1 power unit with only a subtle alteration in cylinder dimensions and would have very similar valve gear and cylinder head arrangements, while from Walter Hayes' departmental standpoint there was something particularly attractive in a story which justifiably could point to the fact that a grand prix V8 measured twice-Cortina, and that there was an intermediate Formula 2 engine to strengthen the link between the two.

However, before any such story could be written, or even hinted at, the grand prix engine project had to be submitted to and passed by Ford's Policy Committee, which is comprised of a dozen or more of the company's top executives who meet regularly to take all the important decisions relative to the operation of the company.

The meeting in question took place a bare six days after Stanley Gillen had arrived in Britain from the United States as Ford of Britain's new Managing Director, and he must have been just a little surprised when, after running smoothly through his first agenda, he turned finally to Any Other Business and heard Walter Hayes say, 'Yes, Harley and I would like to do a grand prix engine'. Hayes then expanded a little, outlining the proposal and explaining that he had had long and detailed discussions with Harley Copp, who had agreed that it would be a good thing. Then, after a short discussion during which Copp confirmed his support, the proposal was put to a vote and passed; it was as simple and as unmelodramatic as that. That a project guaranteed to cost the company a minimum of £100,000, and involving it in activity of which it had virtually no first-hand experience, should be sanctioned with such a lack of red tape was conclusive evidence of, and a considerable tribute to, Ford's corporate policy of relying upon the judgment of its own experts. Walter Hayes and Harley Copp had sold the Policy Committee an idea; now it was up to them to see that it worked. A few weeks later, on a routine visit to Detroit, Hayes received equally emphatic encouragement from Henry Ford II, who has since followed the progress of the project with close and interested attention.

One of Hayes' first actions was to ask Ford's legal department to draw up a draft contract in the form of a tripartite agreement involving the Ford Motor Company, Cosworth Engineering and Keith Duckworth but Duckworth was not to sign it until several months after he had started work on the project.

Keith's reluctance was motivated by his desire first of all to see how things progressed on the design side; although he was excited and stimulated by the challenge now offered to him, his natural caution had allowed him to nurse one or two reservations regarding his ability to see the job through successfully. Consequently he preferred to remain unfettered by a contract, even one containing a clause explaining that his commitment to produce a Formula 1 engine would only apply subject to the successful completion of the Cortina-based Formula 2 engine, for which £25,000 of the contract sum was to be allocated.

The agreement, which became effective on March 1, 1966, but was not signed until June 23 that year, included the following clauses:

1. In consideration of the payment by Ford to Cosworth of the sum of £100,000 . . . Cosworth and Duckworth agree to design

and develop for Ford Formula II and Formula I engines to be known as 'Ford' engines . . . suitable for racing in international competitive events.

2. That the engines to which this Agreement refers are (a) a 1,600 c.c. four-cylinder inline engine designed for the 1967 International Formula II category designated the FVA using the Ford 120E cylinder block and a light alloy four-valve-per-cylinder head, and (b) a 90 degree V8 3,000 c.c. engine using cylinder heads basically the same as/or developed from the above FVA suitable to race in International Formula I races under the regulations which started in 1966.
3. Cosworth and Duckworth will develop the FVA engine during 1966 and produce at least five engines for racing during 1967.
4. Cosworth and Duckworth will design and build and develop the Formula I engine and have engines ready for use by May 1967. They will build at least five engines by January 1, 1968, and maintain such engines until December 31, 1968. Cosworth will supply to Ford a total of five engines suitable for Formula I by January 1, 1968 for the purpose of equipping one Formula I team. The choice of team will be at Ford's discretion, Cosworth being available in an advisory capacity if required.

Chapman's reluctance to put pen to paper on an engine contract offered to him was for a different reason. In his case he wanted certain longer-term commitments written into the document concerning exclusive use of the Formula 1 engine which Ford at that time felt they were unable to undertake. Consequently, Walter Hayes wrote him a letter of intent in lieu of the formal contract, outlining the basic terms of the agreement which had been drafted by Ford's legal department.

Walter Hayes had been a little worried by the rather arbitrary way in which Keith Duckworth had costed the project. Duckworth had been promised the full support of Ford's large research, foundry, engine and laboratory resources, but it struck him that the figure of £100,000 was more an inspired guess than an accurate assessment of the cost of the time, services, materials and equipment which would be required to complete the job. Accordingly, he told Duckworth that, provided the Formula 2 engine was successful, if he found that he had spent more than the anticipated

£100,000 before the completion of the Formula 1 part of the agreement, the Ford Motor Company would be prepared to reimburse him for any reasonable additional expenses he incurred in doing so.

In fact the company did incur additional costs over and above the original £100,000 as a result of their continuing involvement in grand prix racing, but much of the additional expenditure went on the supply and maintenance of a certain number of engines for specific teams or organizations. The original estimate of £100,000 for the design, construction and development of the initial batch of engines proved to be surprisingly accurate, which demonstrated convincingly that provided the brainpower and the will was forthcoming it was not an excessively expensive matter in 1965 to produce a successful grand prix racing engine. Indeed, the subsequent success of the DFV engine must have convinced the Ford Motor Company that at £100,000 it was the bargain of the decade, for such are the expenses of operating a vast industrial complex that it had cost them no less than ten times as much merely to put synchromesh on the bottom gear of the Cortina!

3

Four remarkable men

IT IS WELL KNOWN in motor racing circles that the name Cosworth is an amalgamation of parts of the surnames of Mike Costin and Keith Duckworth, two former Lotus employees who, in 1958, decided to go it alone and form themselves into a light engineering company of service to the racing world. Along the way they inherited two other partners, Bill Brown and Benny Rood, and all four people have played their own very significant role in building Cosworth Engineering Limited from a modest tuning shop in North London into one of the fastest-expanding companies in Northampton where, at the end of 1970, they had 23,000 square feet of factory space and plans to double it during 1971, a payroll of 130, an estimated £400,000 of capital assets, all fully paid for, a turnover in excess of £750,000 per year, and an almost embarrassing amount of liquidity.

Theirs is one of the great success stories of the motor racing world, yet it is a story which hitherto has received little recognition beyond their close associates because the four co-directors have deliberately shunned publicity, both for themselves and for their company. Collectively, they have an almost pathological distate for 'bull' in all its forms and manifestations, and one would be tempted to say that they carry their self-effacement too far were it not for the fact that they have achieved such remarkable commercial as well as personal success.

Yet Duckworth, Costin, Brown and Rood are anything but introverts. Whereas so many engineers tend to be talented and dedicated but rather dull individuals, the Cosworth quartet are extremely amusing as well as unusually clever people, sharing a rich sense of humour, an eloquent gift of repartee and a passionate interest in enjoying themselves in both work and play; their hobbies include flying, gliding, water-skiing and power boat racing, in all of which they excel. Questioned about the talent at the top of Cosworth Engineering, Keith Duckworth has gone on record as saying of the four-man team, 'I suppose we are all a little above average in various ways'. This is a typical Duckworth understatement of fact. The truth is that they are four very remarkable men.

Although Duckworth and Costin formed the original Cosworth partnership, Keith's first contact with his future business colleagues was with Bill Brown, now in charge of the company's sales and general administration. It happened on the day he returned from a vacation at his home in Blackburn to his twin-bedded room in London, where he was studying for a degree in mechanical engineering at the City and Guilds College of University College, London University. He opened the door of his room to find Brown sitting on the other bed.

Duckworth had returned to London with an old friend, who wasn't particularly impressed with the newcomer's lack of conversation when he met him in the room, and he ventured to suggest to Keith when they were alone later that he might have landed himself a pretty depressing sort of roommate.

But that evening Duckworth invited Brown out for a few drinks, figuring that as a mature 20-year-old with experience of the 'Big City' he would be able to contribute something to the social education of the newcomer, who was a mere 19 years of age, and obviously quite raw into the bargain. But by the time they had returned to their room Duckworth's opinion of his new-found roommate had risen by leaps and bounds, for the 'raw recruit' had succeeded in drinking him under the table! Whereas Duckworth felt terrible the following morning, Brown was scarcely the worse for wear. 'I was in pretty good training at the time', he recalls, modestly.

From that day the two undergraduates became firm friends, and they engaged in numerous escapades together, several of which of necessity have remained unchronicled. Duckworth's transport at that time was an Austin Seven Ruby saloon, a device which, like so many of its kind, was inclined towards erratic braking, possibly because the brakes seemed to function only on opposite corners of the car. On one memorable occasion he was escorting his future wife, Ursula, down London's Regent Street, a section of which at that time was paved with wood blocks. Suddenly the traffic lights changed to red, and Duckworth braked hard. As he did so the tail swung around, so he released the brakes and tried to straighten the car, then braked again. But the same thing happened, only more violently, and this time the car did a complete about-face, and went through the red lights travelling smartly in reverse. Ursula thought the episode was much less amusing than her boy friend, who roared with laughter as he eventually brought the car to a stop, then drove forwards on to the pavement in order to get sufficient lock to turn the car back into the required direction.

No doubt it was with considerable relief that she learnt subsequently that the car had been sold to Bill Brown, but Duckworth's troubles with the Ruby were not yet over. Shortly after acquiring the Austin, but before the log book had been transferred to his name, Brown was stopped by the police, who had been intrigued by the car's appearance and were convinced that it contravened a considerable number of regulations governing the use of cars on the public highway.

Brown cannot recall the original reason for the police's interest, although it could well have been the car's lack of front wings. But this was only the start, and the police had an absolute harvest as they examined the device with growing disbelief. In the end, Duckworth, who was still the legal owner, had a summons listing no less than 18 offences, amongst the most bizarre of which was one recording the fact that the car had no gear lever (a tubular spanner had been placed over the selector rod protruding from the top of the gearbox), and another which mentioned the fact that the highway could be seen from inside the car (by looking down at the floor rather than through the windows).

Even its new owner was by now convinced that the car was in need of some attention, and so, like all true motor sports enthusiasts, he decided to convert it into a special, having first seen it survive the long and at times entertaining journey from University College, London, to his parents' home in Hartlepool, County Durham. Brown's car became a popular topic of conversation at the University, but of even greater fame was the prowess of its driver, who found it virtually impossible to steer and propel the thing backwards without hitting something – usually it was something fairly immovable like a wall, but frequently it was someone else's car.

Duckworth's and Brown's residence together in London was cut short after a while when Brown's tutors confirmed their worst fears regarding his academic prowess. Though he worked incredibly hard on the social side of life, he put less effort into his engineering studies, and on failing a set of examinations he was invited to leave. He returned subsequently for another crack, but with similar unfortunate results, and so ventured forth to seek fame and fortune without the backing of any academic status.

But despite the lack of letters after his name, he was still a sound, practical engineer, and after joining a chemical engineering works in his home town he soon found that he was virtually in charge of the place. However, his employers thought that perhaps it was tempting fate to entrust their business to an unqualified 21-year-old and so informed him that they would be bringing in an older, qualified man over him, whereupon Brown decided to leave his protected job and joined the Royal Air Force for his National Service.

Keith Duckworth, meanwhile, had moved in the general direction of Lotus. On selling the Ruby to Brown he had invested in a kit of Lotus 6 sports car parts, which he proceeded to assemble during one of his vacations from University, and into which he dropped a Coventry Climax engine, one of the initial batch of 25 of these fire pump engines which found their way into motor sport. To own a Lotus in those days inevitably meant fairly frequent visits to the factory for the purchase of various bits and pieces to replace those which had either dropped off or had been knocked off. It was during one of these visits that someone at Lotus mentioned that they were permanently in need of skilled pairs of hands, whereupon Duckworth decided that working on sports cars might be a more congenial form of vacation training than the tapping of vast threads for gas engines which had been occupying his time of late. And so he became one of the growing band of part-time helpers who contributed so much to the establishment of Lotus in its formative years, following in the footsteps of Mike Costin, whose enthusiasm for motor sport had drawn him towards Colin

Chapman and Lotus several years earlier, first as a voluntary helper, and subsequently as a full-time employee.

Mike Costin's background had been with the De Havilland aircraft company at Hatfield, from which he emerged as a very talented practical engineer and a tireless worker – two ingredients in great demand at Lotus at that time. By the time of Duckworth's arrival on the scene, Costin was fairly high in the Lotus hierarchy, but Keith's immediate boss turned out to be Graham Hill, who at that time was in charge of gearbox assembly, but who was in the habit of taking off from time to time to go motor racing for whoever would use his considerable abilities in that sphere. It was during one of Hill's 'quick vacations' – to Le Mans in this case – that Duckworth's own very considerable talents were brought to the attention of the Lotus management. Until that time Graham Hill had been the only person on the payroll who could assemble the modified Austin A30 gearboxes which the company were using for their sports cars, the general feeling there being that this was a vastly intricate job requiring exceptional skill and experience – a notion which Graham had done his best to maintain by the simple expedient of looking after all this work himself!

But Lotus were in desperate need for more gearboxes, and when Hill took off for France Duckworth was asked whether he thought he could tackle the job. He had never assembled a gearbox before, but he said that he couldn't see why it should be all that difficult because, after all, it was only a kit of parts which presumably went together in some sort of logical order. His simple identification of the problem and his practical solution to it made him something of an overnight hero at Hornsey where the gearbox crisis was soon overcome, although Graham Hill was somewhat displeased on his return from Le Mans to find that his services had proved to be not quite as indispensable as he had imagined! But, in any case, Graham was intending to leave the company fairly soon as his racing activities were developing well, and as this coincided with Keith's successful conclusion of his University degree course, Duckworth was invited to join the full-time staff of Lotus as a development engineer on gearboxes. In 1957 graduate engineers were in great demand, and there was almost a queue of potential employers at the University at the conclusion of the academic year, but Duckworth was attracted by the lure of motor racing more than by the greater financial rewards to be found in less glamorous occupations, and so agreed to join Lotus for a salary of £600 per year.

It was at this point of time that he first came to know Mike Costin really well, not only because they were working closely together but because they would also spend many hours during the evenings just 'talking shop' and benefiting from each other's often opposing viewpoints. They became an ideal combination, for Duckworth was very strong on engineering theory while Costin was equally strong on practical solutions to engineering problems. They obviously taught each other a lot during this period, and both probably emerged with a more balanced approach to their respective jobs.

Duckworth's main activity during this time was the development of Colin Chapman's five-speed Lotus gearbox which was being used on the first Lotus single-seaters as well as on the larger-capacity Coventry Climax-engined sports cars. It was an ingenious piece of mechanism, but it was very troublesome, mainly because it was being called upon to do too much. After a while, Keith became frustrated by the amount of time he felt he was wasting trying to develop the transmission which, in his view, needed a complete redesign. The final straw came when he was told to carry through another detail development programme for the next racing season. He decided that this was just another waste of time (subsequent events were to prove him right), and after quite a long discussion with Colin Chapman he resigned and decided to start his own engineering business.

Duckworth and Costin had been discussing the possibility of joining forces for some time, for it seemed a logical way of exploiting their mutual talents. But although they formed Cosworth Engineering Limited almost immediately, the intention was that initially only Duckworth would work full-time for the company, and that Costin would simply lend a hand during evenings and weekends when his Lotus duties permitted. In fact, Mike was able to contribute virtually nothing for the first three years because shortly after the Cosworth formation Colin Chapman offered him the post of technical director of Lotus, with a three-year contract, one clause of which specifically excluded any extra-mural activities.

Initially Cosworth Engineering had no premises of their own, and

Keith Duckworth started the business by working out of a workshop operated by the racing driver John Campbell-Jones in West London. But eventually premises were found adjacent to a public house in Friern Barnet, North London, where immediately Duckworth installed a dynamometer, having managed to scrape up the collateral for a bank loan of £600. He had been amazed by the lack of dynamometers amongst the various companies purporting to be engaged on engineering for motor racing customers, and could not imagine how these concerns could hope to do their job properly without one.

Soon he was tuning Coventry Climax engines very successfully, as well as carrying out a number of suspension and other chassis conversions on customers' cars, and in a very little time he had built quite a reputation for his young company. As business boomed, helping hands were enlisted, and it was about this time that Bill Brown came back on to the scene. Still in the Royal Air Force, he managed to obtain a most convenient posting which put him in charge of a piece of radar equipment only an hour's drive from Friern Barnet. As this particular piece of equipment was inclined towards unreliability, from time to time Bill would encourage the frequency of its failures, and having ensured that it was not working would take off to spend the remainder of the day helping out his friend and former drinking partner. As the posting also provided him with transport and a driver, the arrangement between Brown and the Government could scarcely have been more convenient, at least for Brown. In at least one respect, therefore, it could be said that the British Government, through its armed services, made at least a token contribution towards the welfare of motor racing during 1959 and 1960!

The early development of Cosworth Engineering coincided with the growth of Formula Junior into a full international racing class, and Duckworth decided to design and build a Cosworth FJ car, which was to be raced by Howard Panton, a friend and customer who was working for Ford of Britain. (Duckworth's own racing career with his Lotus 6 had been comparatively brief, for he had found it an expensive hobby, especially after his final race, during which he had contrived to demolish the chicane at Goodwood and cover his battered car with geraniums!).

Duckworth's intention had been to power his Formula Junior

car by a Fiat 1100 engine, but he was having great difficulty in extracting sufficient power from it when Panton revealed that Ford had a new overhead-valve Anglia engine on the stocks, which should prove far more suitable for FJ racing. Duckworth obtained an early example towards the end of 1959, and began a frantic development programme in the hope of installing it in a car in time for the Boxing Day meeting at Brands Hatch, which was to see the British debut of Formula Junior, the class of single-seater racing having been devised some years earlier in Italy by Count Johnny Lurani.

So much time had been taken up on engine preparation that all work on the Cosworth car had to be abandoned, but after many setbacks and much toiling right through to Christmas morning the Boxing Day deadline was met. In fact there were two Ford engines in the paddock that morning, although one, in Alan Stacey's prototype Lotus 18, ground away its sump and ran its bearings after losing all its oil during practice, and the other, in Graham Warner's Gemini, broke its flywheel, also during practice. As Warner's car was badly damaged he sportingly offered Lotus his engine, and for the remainder of the morning the entire staff of Cosworth Engineering and their helpers were busily engaged in the paddock cannibalizing one serviceable Ford engine from the two and installing it in the Lotus. The job was still being completed as the cars were flagged out on to the circuit, but everything was screwed together just in time, and December 26, 1959 became the date of the first public appearance of a Cosworth Ford engine on a race track.

One of the helpers at Brands Hatch that day was Benny Rood, another man of unusual talents who ran a machine shop in Walthamstow, and who had been doing some sub-contract work for Cosworth. Benny was nearly, but not quite a Cockney, if you accept that a true Cockney is someone who was born within the sound of Bow Bells; in his case, if the wind was in the right direction, the church bells could be heard within a quarter of a mile of his birthplace. But even though strictly speaking he came from 'across the border', Benny inherited the Cockney's great sense of humour, his capacity for hard work when the chips were down, and his abundance of commonsense and ingenuity.

After serving in the Army during the war, he had worked for

his father making packaging machinery for his ice cream and refrigeration businesses, and from 1948 to 1956 he had had a very successful career as a motorcycle racer, scoring many successes in short-circuit races as well as collecting the coveted Governor's Trophy in the 1952 Ulster Grand Prix. Benny used to build his own bikes at the rate of about one per year, until suddenly, when halfway through assembling a new 350 c.c. cycle for the 1956 season, it struck him that if he put half as much effort into turning out bits and pieces for other people as he was expending on his own machines he would have the basis of a good engineering business.

He had already proved himself as a talented intuitive designer, for his motorcycles had shown themselves to be very effective in a highly competitive class, and soon he was conquering a new field as he became involved in the design of a 350 c.c. two-stroke engine for hydroplane racing. The Hoganrood engine was to win its class of the hydroplane championships three years running, from 1957 to 1959, and is still remembered with admiration by people involved in the sport.

Just as Keith Duckworth's early discussions with Mike Costin proved mutually so beneficial, Duckworth's first long talk with Benny Rood, on the return journey from Brands Hatch to London that Boxing Day evening, was also to have a very lasting effect on their relationship. They 'talked shop' all the way home, and Duckworth, who has always been a deep-thinking person, was intrigued with Rood's brand of engineering philosophy. During the early months of 1960, when orders started to pour in for Cosworth Ford Formula Junior engines, Rood's machine shops were used more and more for Cosworth business, and Benny would spend long hours with Duckworth helping to overcome the latest development problem with camshafts, valves, tappets or whatever components were providing the current headache.

Although Mike Costin's absence from the business was unfortunate it had proved highly beneficial in at least one respect for he had been able to convince Colin Chapman that the Cosworth engine was worth considering for the Lotus Formula Junior cars at a time when Chapman was in no great mood to further the advancement of his recent employee. As it transpired, the Lotus contract really set Cosworth on their feet, for Chapman and his staff were selling FJ cars like the proverbial hot cakes, and they were to take 126 Cosworth Ford engines that year, all of which were delivered to Friern Barnet in standard production form and returned to Lotus in full race trim with a guaranteed minimum output of 75 b.h.p. for the unbelievably small sum of £145 apiece.

Meanwhile, Bill Brown had concluded his strenuous period of service with the RAF, and from the end of 1960 his visits to the Cosworth factory became legitimate as he joined the staff as general business and sales manager. This move was something of a disappointment to his father, who had been academically brilliant but unable to enjoy the University education which he desired so much, and who had been hoping that Bill, having been given two chances, would now return for a third attempt to obtain a degree as a passport to a worthwhile career. It was some considerable time before Bill was able to convince his family that he had done the right thing.

It was not until July 1962 that Mike Costin concluded his Lotus contract and became free to join Cosworth Engineering, but throughout the three intervening years he had kept strictly to the terms of his agreement with Colin Chapman, and had confined his meetings with his partner to occasional Sunday afternoon chats. Nevertheless, he still recalls the pleasure he felt when, on the day after he finished at Lotus, he returned there with Keith Duckworth to talk to Chapman in his capacity as one of Cosworth's customers! In this instance the topic of conversation was the contract for the further development of the Twin-Cam engine which was to be fitted to the proposed Lotus version of the Cortina saloon, final assembly of which was to be carried out at the Lotus factory at Cheshunt on behalf of the Ford Motor Company.

By this time Cosworth Engineering had outgrown their premises at Friern Barnet and moved into much more modern facilities at Edmonton, utilizing the very building in which the prototype Lotus Elite had been built in the utmost secrecy. Benny Rood's machine shop was now producing work for Cosworth virtually to the exclusion of all other business, and for this reason certain Cosworth machinery was installed at his premises at Walthamstow. By this time Benny was so busy that his paper work began to suffer, and there was a general air of confusion surrounding the finances of the liaison between the two companies which eventually was resolved by mutual agreement whereby Cosworth hired the full services of Benny's company for a fixed sum per week, to include the payment of staff and materials, while Benny himself remained self-employed, but also engaged 100 per cent on Lotus work. This arrangement enabled Cosworth staff to be moved from Edmonton to Walthamstow as necessary, and eventually the Walthamstow premises became in effect the Cosworth machine shop, which relieved the pressure on space at Edmonton until such time as the company could make their big move and expansion at Northampton. Later, Benny was offered a seat on the Cosworth board as the final act in the merger of the two companies.

Throughout the Sixties, Cosworth were to go from success to success. The Formula Junior engines, of which there were to be several derivatives, all based on the Ford Anglia power unit, gave way to a basically similar Formula 3 engine for 1964, and coincident with this, the first genuine Cosworth, as distinct from Cosworth-converted, racing engine, the Formula 2 SCA.

After a while, Cosworth discontinued Formula 3 engine manufacture as they concentrated on their Formula 2 and Twin-Cam programmes and it was a logical progression that they should choose to develop their FVA 1·6 litre Formula 2 engine for the 1967 formula from the latest Cortina cylinder block, thereby providing the foundation not only for the magnificently successful DFV Formula 1 engine, but also for a 1·8 litre four-cylinder FVC engine for sports car racing.

Such has been their growth that Cosworth, who moved to Northampton (because it was within five miles of the M1 motorway and easily accessible to both London and the Midlands) in 1966 and built 6,800 square feet of office and factory space plus 720 square feet of dynamometer shops, were obliged to add a further 3,000 square feet within 18 months of moving in, and no less than 12,200 square feet on top of that after a further 18 months. Then, at the beginning of 1971, they finalized plans for yet another 22,000 square feet of factory and office space on land acquired on the opposite side of St. James' Mill Road, where they have become a local landmark.

They even have their own heliport, where Keith Duckworth parks his pride and joy, his two-seater Brantley 'chopper'. At £14,000 it might have seemed something of an indulgence when he acquired it in 1967, but as the cheapest helicopter buyable in Britain in 1971 cost in the region of £23,000, it proved itself to be a very sound investment over those four years.

Duckworth's choice of a helicopter, in preference to the more conventional type of aircraft favoured by so many motor racing people, was made purely on practical grounds. He liked the idea of private flying, and he wanted to combine a pleasureable hobby with a business convenience and asset. He figured that in a country as congested as Britain, the helicopter represented the only practical form of air travel. The 'chopper' is said to be a much more difficult device to fly than a conventional winged plane, but Keith bought the Brantley before he learnt to fly it. It took him only 37 flying hours, which is yet another rather remarkable achievement.

His other hobby is water-skiing, although he says that he also puts thinking high on his list of relaxations. This is no facetious remark; he derives great enjoyment out of original thought, not merely on engineering matters but on any subject whatsoever. He even finds pleasure out of contemplating privately the great problems of society, and his pleasure seems in no way dimmed by his conviction that the vast majority of them are insoluble! It is, perhaps, his unusually acute powers of original thought which cause him to choose his words for conversation with great care and to deliver them with measured emphasis. He emerges as quite a philosopher, with an ability to serve up a most palatable dish of home truths and fundamental wisdom dressed liberally and tastefully with good humour.

Some of this, at least, has rubbed off on his colleagues. Bill Brown, for instance, whose physical resemblance to Duckworth is such that the two are often confused, is another eloquent and articulate conversationalist, and the member of the quartet who more than any of the others is bitterly opposed to anything even vaguely classifiable as 'bull'. It is a condition imposed not through shyness but by down-to-earth honesty. As the business administrator of the company he operates an ultra-democratic organization, where not only are christian names used throughout, but where top management share the same facilities with the newest and youngest unskilled employee, right down to pouring their cups of tea out of the same large metal pot. Bill will admit that perhaps he overdoes the democracy at times, but he cannot conceive of any other way. It certainly seems to work.

He tends to make little of his own very considerable contribution to the successful growth and financial stability of the company, and is more inclined to recall the fact that he spent too much time 'chasing the birds and drinking beer' at college when he should have had his head in books. Yet if pressed he will admit, a little reluctantly perhaps, that he considers himself as someone with an engineering background who tends to think of things in engineering terms. Press him even further and you will find that he has also involved himself in a measure of designing, 'although I doubt whether Keith would consider it design!', and that some of the Cosworth components in the earlier days of the company were produced from his drawings.

If Bill differs from his colleagues in any respect it is that while they have taken to the air he finds his sport on or beside the water, either racing power boats or the other extreme, fishing. It was Benny Rood's hydroplane connections which first attracted Brown to power boats, in which subsequently he has excelled as a competitor. He met Benny's friend Roger Hook in 1960 and became his co-driver for a number of seasons before building and racing his own boats which he calls Venus. Understandably, he has used an engine – the Ford Twin-Cam – prepared by Cosworth, which puts him in the R3 class up to 2 litres, and he won the international championship three years running from 1966 to 1968. The following year he achieved fewer results, and in 1970 he was forced to abandon racing because of two family bereavements. However, he is closely involved with the sport's administration as a member of both the Sports Boat Committee and the Technical Committee of the Royal Yacht Association, and during the winter of 1970/1971 was preparing to break new ground by building the first twin-hull catamaran with an inboard engine. He was also preparing himself to be petrified by its anticipated top speed in the region of 120 m.p.h.!

Mike Costin, physically the giant of the quartet, and the person who steers Duckworth's designs through the prototype stage and into production, is another Cosworth man who is no stranger to high speed. He is one of those rare mortals who, despite driving racing cars comparatively infrequently, has been able to climb straight in and drive them to victory time after time. This ability has proved extremely useful on occasions, notably when, during the initial development programme for the FVA engine, he was able to test it under race conditions in a secondhand Brabham chassis which the company purchased expressly for this purpose. In this way, the prototype engine was fully developed before the power unit was put into series production for customers, who subsequently have been grateful for its outstanding reliability.

Mike also had the considerable honour of being the first person to drive the Ford DFV-engined Lotus 49, initially on the Lotus test track at Hethel, and the following day at Snetterton while both Jim Clark and Graham Hill were abroad. He was duly impressed, not only by the tremendous performance of the engine on its first visit to a race circuit, but also by the somewhat alarming antics of the Lotus, the reason for which was traced to a pair of rear wheels which were toeing-out several degrees! One way and another Mike Costin had quite a day, for he had borrowed Colin Chapman's plane for the journey to Snetterton, and on the return trip was staggered to come across an RAF plane parked at 2,500 feet! He couldn't believe his eyes, but in fact he was looking at the very first and still very secret Harrier vtol jump jet; perhaps it was as well that he resisted the temptation to lower the landing gear, shut down his engine, and step out!

Power flying occupied much of Costin's spare time for some years before he switched his main interest to gliding, a sport in which he quickly became highly proficient; using a Cirrus glider he was sharing with Benny Rood and another friend, he took fifth place in the national gliding championships in 1970. Costin has always been a considerable athlete, and in particular he shares Keith Duckworth's keen interest in water-skiing.

Benny Rood has been power flying and gliding since 1966, but the man who used to race around the Ulster motorcycle road-race course at speeds of well over 100 m.p.h. is also keen these days on more modest and gentler horsepower, and can count in his transport stable a horse and a pony, the latter being used mainly by Mike Costin's children. He is, perhaps, the least well-known of the Cosworth bosses, because his activities have tended to keep him close to base at Northampton, when one or two of the others have been in evidence at race meetings and other motor sports functions. But his contribution to the growth of Cosworth has been very considerable, and as the person most closely involved with the administration of the production side of the company his is a vitally important role, and it is to his great credit that Cosworth have been able to provide such a fine service to their customers, sometimes in conditions of great stress.

These, then, are the four men who steer Cosworth Engineering. Their duties are not easily defined because they work as a team, and that means a high degree of overlap. They have a great deal in common – what Keith Duckworth might call 'above average', but most other people consider to be outstanding ability; a capacity for and an enjoyment of hard work; sound business as well as engineering judgment; a love of and an ability for high-action sport; and above all a sense of purpose coupled with an inherent honesty which has been fundamental to their continuing prosperity. They are, indeed, four very remarkable men.

4

Duckworth's design and philosophy

IF KEITH DUCKWORTH'S Ford DFV engine were to appear in a catalogue of racing car engines its entry would probably read something like this:

Ford DFV engine for Formula 1 racing. 90-degree V8, water-cooled. Bore 3·373 in.; stroke 2·550 in.; capacity 2,993 c.c. Compression ratio 11 to 1. Maximum power approximately 410 b.h.p. at 9,000 r.p.m., rising to approximately 430 b.h.p. at 10,000 r.p.m. in 1970. Maximum torque 245 lb ft at 8,500 r.p.m.. Lucas high-pressure fuel injection into inlet ports and Lucas transistorized ignition. Cylinder block cast in heat-treated aluminium alloy with cast-iron wet cylinder liners. Heat-treated aluminium alloy cylinder heads with two gear-driven overhead camshafts to each cylinder bank. Two 1·32 in. diameter inlet valves and two 1·4 in. diameter exhaust valves per cylinder with 32-degree included angle and 0·41 in. lift. Inlet valve opens 102 degrees ATDC. Hepolite pistons to Cosworth design, Vandervell lead indium 1·937 in. diameter big-end and 2·375 in. diameter main bearings and nitrided steel crankshaft. Autolite PG 403 plugs. Weight 356 pounds.

But these would have been merely the bare facts, and although such a specification suggests an unusually efficient design it gives no indication of the DFV's characteristics, its personality, or its success. It does not explain that perhaps a dozen different people could have designed such an engine to the same basic specification, but failed to produce anything approaching the calibre of Duckworth's DFV.

One of the fundamental reasons for the DFV's success is that

Duckworth's basic design philosophy was right. Furthermore, having decided (correctly) what was needed, he knew how to design the engine in such a manner that it could be manufactured correctly and maintained efficiently, and so produce – and keep on producing – its design performance.

It is unreasonable, even naïve, to expect a successful designer to reveal the innermost secrets of an engine which has won the vast majority of races for which it has been entered and which is still current. Nevertheless, Duckworth has been prepared to talk at considerable length not only about the design of the DFV and the method by which he arrived at certain decisions, but also about his design philosophy.

Attempts can be made to interpret a designer's thoughts and considerations and to evaluate his aims and his methods of achieving them, but in the final analysis you can only get beneath the skin of a design by listening to the designer himself.

The remainder of this chapter, therefore, is written in Keith Duckworth's own words, as he spoke them to David Phipps.

'My approach to design is very fundamental. I try to start with no pre-conceived ideas and with an open mind, and faced with a problem I attempt to analyse the requirement. In the design of racing engines there are very few really firm facts. However, there are a lot of probabilities, and in designing a cylinder head – which is the most important part of an engine as far as power output is concerned – it is necessary to consider all the factors which seem to be relevant. These include obvious things like breathing and combustion, as well as inertia stresses and other things which would affect the bore and stroke. Having taken all of these into consideration it is necessary to make a qualitative guess as to what seems to have the best chance of succeeding. It is then necessary to go into detail on the concept of combustion chamber design, cylinder head design and bore and stroke, which gives a reasonable compromise solution. I don't think this can be done quantitatively. I think that in order to obtain a satisfactory result it is necessary to have either flair or a reasonable ability to judge probabilities. I start by designing the cylinder head, and then try to make everything else fit in with it.

The mathematics of the loads involved in engines are so involved, with the effect of random vibrations, torsionals and other things creeping in, that to stress from calculated loads is open to very great errors. Past experience, plus an eye for avoiding stress concentrations, is much more likely to provide a satisfactory result. And even if you haven't a great deal of personal experience, as I hadn't when I started on the DFV, it is possible to learn quite a lot by looking at other people's connecting rods and other people's crankshafts.

On the question of cylinder head layout, I have never been very much influenced by other people's design. When I produced the SCA engine it was said that it had a Heron head. At the time I did not know that Heron had produced this configuration, and the only reason that I arrived at it was that I was used to working with bath-tub combustion chambers. At the compression ratios that it seemed worth running, the valve recesses provided all the combustion chamber that was required. By putting the clearance slots in the piston – rather than recessing the valves into the head – I avoided masking the valves, because the piston was out of the way for most of the stroke. The fact that somebody else had produced a flat-headed engine, and his reasons for doing it, were completely unknown to me.

When we started work on the DFV, everybody else seemed to think that 12 cylinders was the absolute minimum for a 3 litre Formula 1 engine. BRM had their H16, and there was even talk of somebody going to 24 cylinders. We didn't really consider making anything other than a V8, and even now, five years later, I'm still not convinced that more cylinders are necessary. All the existing 12-cylinder engines are being run considerably faster than the DFV to give similar power outputs, which nullifies the advantage of their shorter stroke. Their maximum piston accelerations are probably at least as high as ours, and therefore their problems in increasing their r.p.m. are just as great as ours.

As we weren't involved in the 1½-litre formula we didn't learn a great deal from it, and in practice our 1,600 c.c. four-cylinder engine had a higher output than the 1,500 c.c. eights of that era. Whereas everybody thought that eight-cylinder engines would be required to get power outputs approaching 150 b.h.p. per litre, we achieved this using a four-cylinder.

From consideration of the valve area that would be required, it was obvious from our experience with the 1,000 c.c. SCA engine that we would have to use four valves per cylinder. At that stage the merits of four valves per cylinder were not as clear as they are now; BRM had tried four valves with unsatisfactory results, and Coventry Climax had only made a small gain. Only Honda had really proved the advantages of four valves, and this was because they were the only people who had managed to get their detail work right; in itself the concept has always been sound, but only if the execution is up to the same standard as with a good two-valve layout.

When we started on the DFV the single-camshaft Repco V8 was winning most of the races, but it seemed unlikely that this situation would continue as their success had been due only to the unpreparedness of the opposition. The BRM H16 was extremely complicated and was prone to extraordinary torsional problems. When I heard about the gearing together of the crankshafts and the way the valves were all driven from one crank I felt that a basic error in layout had been made. The basic configuration obviously had attractions, but the complications and difficulties of it really frightened me. The Maserati V12 seemed to have been developed from various older engines and had a rather untidy layout, and although Ferrari were experimenting with three valves on their V12, I didn't feel that a layout of this type had a great deal of merit.

Potentially the best engine at that time was the Eagle V12, but it was handicapped by the manner in which it was made – virtually every item in it was a one-off. Ironically it won its only Grand Prix at Spa – the last circuit on which I would have expected it to hold together, and much the same can be said of the BRM V12 which won its first race at Spa in 1970. The V12 Honda should have had a lot of power, but only just managed to beat the Repco at Monza in 1967 and was nothing like as fast as the Lotus-Fords. (Clark made up a lap on it following a tyre change, but then ran out of fuel while well in the lead.)

As I have already indicated, when I think of a new engine, I think first of the cylinder head, and the first head I designed was for Formula Junior, which subsequently developed into Formula 3. We eventually managed to get 100 b.h.p. per litre out of a down-draught push-rod Formula 3 engine, and we went on from this to the 1,000 c.c. Formula 2 engine with the standard Ford Anglia bore and stroke and our own single-overhead-camshaft cylinder head. We tried very hard for three years to get that engine to burn. The volumetric efficiency was very high, but judging from the 49-degree spark advance we needed we had a combustion problem and we never achieved very high b.m.e.p.'s.

Then the capacity limit for Formula 2 changed to 1,600 c.c., and having settled on the four-cylinder Ford Cortina block because there were no suitable six-cylinder units available, it was obviously necessary to get a considerably larger valve area than on the SCA. As the bore could not be increased very much, because of the thickness of the cylinder walls in the block, this meant that if we were to get adequate valve area we would have to make a four-valve engine. At this time the only really successful four-valve engine was the Honda, so we didn't have a great deal of engineering precedent to follow. On the SCA we had attempted to control the combustion by squish, but I felt this was only really useful for tidying up the remote parts of the chamber; in general it occurred too late to be a major combustion control. We did eventually get 143 b.h.p. per litre out of it, but it was obvious that Honda were getting over 150 b.h.p. per litre with their four-valve head.

Having thought about the combustion chamber area and other factors, I felt it was entirely feasible to aim for 200 b.h.p. from a 16-valve, 1,600 c.c. four-cylinder engine. Initially I planned to use the layout subsequently adopted by BMW, a pent roof with diametrically opposed inlet and exhaust valves, but I abandoned this because it would have needed three spark plugs per cylinder, and because it would have been very difficult to arrange the inlet and exhaust systems on a V8 version.

Thus the layout adopted for Formula 2 was extremely simple, with the inlet valves on one side and the exhausts on the other, the angle between them being a relatively narrow 40 degrees. The two overhead camshafts were driven by a train of gears from the front of the crankshaft, and the standard Ford camshaft was retained to drive the water pump.

At that stage we had never designed an engine from scratch. We had made cylinder heads, pistons and connecting rods, but we hadn't made a complete engine. It is one thing to fit special heads to an existing block, with existing studs and existing water passages, but an entirely separate matter to start off with a clean sheet of paper; you may not be at all happy with the existing studs or the existing water passages, but at least they give you something to go on.

I don't want to give away too much of our philosophy on cylinder head design, because it is obvious that most people are just not thinking about the problem. Most of them get absolutely mesmerized by gas flow, mainly because it is something which can be measured fairly easily. The unfortunate thing is that there is no point in getting an awful lot of mixture into an engine unless you can manage to burn it at the right time, and if you try too hard to get high gas flow figures you can easily wind up with a combustion chamber which doesn't combust. I have never believed that there is any point in having a gas flow rig and measuring the flow. I think it is possible to look at the shape of a hole and decide whether air would like to go through it or not. A hole that looks nice and smooth and has no projections will generally flow easily. Most people start with something so horrible that to create an improvement should be very simple. I would claim that I could arrive at something very close to their results from gas-flowing just by putting my finger down the hole and seeing what it feels like; in any case, to be totally mesmerized by the air situation is to lose sight of the object of the operation and is a bit short-sighted.

Having given up the idea of using diametrically-opposed valves, and settled for a conventional pent roof layout, I managed to convince myself that I had a system which was likely to burn most of the air that I could get in; this was with a reasonably compact 400 c.c. chamber – 400 c.c. per cylinder. When it came to making a Formula 1 engine we had to reduce the cylinder size to 375 c.c. To do this we had to decrease either the bore or the stroke, and as the stroke was relatively long on the FVA, having been dictated by the amount we could bore the cylinder block, we decided to shorten this rather than reduce the bore. This necessitated keeping the pistons flat (apart from small valve cutouts) and decreasing the valve angle. At this stage the Formula 2 engine was giving very good b.m.e.p.'s, which suggested that the combustion was not too bad, and so it was with some trepidation that I reduced the valve angle. However, there was an additional advantage in that this made the cylinder head slightly more compact, and as the engine looked in danger of growing very wide this was a consideration.

The overall design of the Formula 1 engine was dictated by the decision to make a 90-degree V8. It is very difficult to construct a chassis of any merit around this type of engine due to the plan area of the cylinder heads, and as a result it seemed logical to use the engine as a stressed member, and to carry the suspension on the engine or the gearbox or on a sub-frame attached to them. This made it necessary to choose some suitable pick-up points on the engine and stress it in such a manner that it would take the usual chassis loads through it. Looking at the engine from the front, the bottom corners of the sump and the two rocker covers provided a circle of pick-ups with a suitable configuration for the back bulkhead of a monocoque, so the decision was made to pick up the engine from these points. As the engine would expand while working we had to be careful that we didn't put stretch into the rear bulkhead, so all the shear was taken at the bottom mountings and two fairly slender plates were inserted between the cam covers and the monocoque; these would allow the engine to expand 15 thou or so without significant stress.

Because of the height of the exhaust pipes and the width of the heads, it was obvious that it would not be possible to carry much fuel alongside the engine. Consequently it was decided to make the engine as short as possible so that the front end could go on to a flat panel without wasting any space. Thus all the accessories that normally are mounted on the front of engines, such as water pumps, oil pumps and distributors, were fitted low down on the side of the engine in the space below the exhaust pipes which would otherwise have been wasted. With such a short engine it was possible, without pushing the driver too far forward, to fit a fuel tank between the driver and the engine, which is the ideal place for it. By carrying the fuel in the middle of the car it is possible to maintain the same weight distribution whatever the fuel load.

I felt that the mechanical efficiency of an eight-cylinder engine would be better than that of a twelve, and that it would be easier to keep the weight down. Better mechanical efficiency would lead to better fuel consumption, and I knew from our four-cylinder experience that the exhaust system would work reasonably if we used a flat crank. I was not entirely happy about using a flat crank on such a relatively large engine, but when we worked out all the sums as to the amount of out of balance it seemed that the maximum transverse shake would be plus or minus 3 thou; we felt that this should not be too serious, even if the engine was rigidly coupled to a monocoque, and it would certainly simplify the exhaust system layout. In practice there have been slight vibration problems on certain cars, mainly through the odd panel being in natural resonance with the engine over a fairly small section of the r.p.m. range. Also there has been a certain amount of instrument trouble through the dashboard shaking the instruments around, but in general the vibration level is acceptable.

It was obvious that for lightness we would have to use an alloy block, and we would have to have some form of liners, the development of high silicone alloy being insufficiently advanced at that time to allow us to dispense with them. The bore spacing that is required with wet liners is considerably greater than is necessary with a cast-iron block, and this might have encouraged us to go for a smaller bore than on the FVA in the interests of overall length, but in the end we decided to use the same bore as the FVA and shorten the stroke from 2·72 to 2·55 inches; the length of the con-rod was increased to avoid an unacceptable increase in piston acceleration.

I have always felt that mechanical losses are significant in engines, and on the DFV I made a special effort to keep the aerodynamics of the crankcase reasonable. If the crankshaft bob-weights are travelling around at 170 m.p.h. it seems likely that the air in the crank chambers will be revolving with them. Thus I tried to keep the internal surfaces of the crankcase as smooth as possible so that the air could pass over them easily. I was also very much aware of the problems of getting oil out of the crank chamber, and anxious to get the centre of gravity of the engine as low as possible. For this reason a very small flywheel was chosen, and the sump was designed to get the crankshaft as close as possible to the ground. The result was a fairly complicated bottom half, with the main bearing caps integral with the sump; this also made it structurally compatible, so that chassis loads could pass through the bottom of the sump. We also had a fairly unusual oil system.

As far as liners were concerned we adopted an ordinary iron type, sealed by two 'O' rings at the bottom and Coopers mechanical joints at the top. We mounted the water pumps on the sides of the block at the front end, with the oil pumps behind them. The cam carriers were made very rigid, so that chassis loads could go through them, on the way from the cam covers into the head, and thence into the block without any risk of tappet seizure; it was also thought that all these joints should be made metal-to-metal; with paper gaskets the reduction in friction in the joints might have led to chassis loads disturbing engine components.

Because we were using a very small flywheel, and because we wanted to get the engine as low as possible, we had to adopt a special tooth form for the starter ring, and we had to mount the starter motor alongside the gearbox. (Because the flywheel was so small it was impossible to get a starter motor alongside the engine that would mesh with this starter ring.) It was also felt to be a good thing to move the weight of the starter motor as far back as possible, and the gearbox side provided an excellent mounting for it. (I have always been unhappy about cantilevering something as heavy as a starter motor out from a flange.) A conventional V8 configuration was chosen, with the left-hand bank forward and the right-hand back, to avoid the use of forked con-rods, which I regard as extraordinarily complicated and not worthwhile.

As far as materials were concerned, all the major castings were made in LM 8 WP heat-treated aluminium alloy, a good crisp alloy which is fairly easy to cast and has very good machining properties. The cover castings were made of magnesium alloy, and the crankshaft was of conventional nitrided M 40B – the same material we had used in all our previous engines. The big-end bearing diameters were the same as on the Formula 2 engine, which was the same as on the Ford four-cylinder range though narrowed down to minimize the block offset because the area was adjudged to be unnecessarily large on the standard Ford.

The main bearing diameter was enlarged, to increase the overlap between the pins and the mains, and to increase crankshaft stiffness. A fair proportion of the rotating and reciprocating weight was balanced out to minimize the main bearing loads, but unfortunately this dropped the natural frequency. It might have been better to increase the main bearing loads and decrease the weight to put up the natural frequency; as it is we have been passing through it continually at 86, with a secondary resonance at 96.

The design of the pistons and con-rods followed the same lines as on the Formula 2 engine. The gear train at the front of the engine incorporated the quill in the nose of the crankshaft that we'd found essential first of all on the SCA, and a train of 14 gears took the drive up to the camshafts. We found we could just get an integrated unit into the Vee comprising a small capacity alternator, a fuel injection pump, the fuel injection metering unit and the distributor – provided we didn't use the whole of the distributor. We had used Lucas fuel injection on the Formula 2 engine and the system had always appealed to me as being one that would work, so we felt that the eight-cylinder Mark I Lucas fuel injection system was a sound bet. The ignition equipment which was proposed by Lucas was Opus, which was similar to the system we had used on the SCA, and was, in fact, an eight-cylinder version which was being developed for use in a production car.

Because we were trying to get it all inside the Vee, in between the induction manifolds, we had to take two parts of the Lucas distributor, the cap and the base, and drive them with a shaft of our own design; this naturally has caused any trouble on the ignition side to be blamed on us, because Lucas couldn't assemble the distributor complete and test it. There was originally a suggestion that the engine should be fitted with Autolite ignition, but this idea was dropped because of the very high development costs involved.

At the time when we were designing the engine a lot of cars were having fuel vaporization problems, which seemed to be due to excess fuel picking up a lot of engine heat before being pumped back to the tanks. To obviate this we decided to pump the fuel around inside the inlet manifold casting, where it would be cooled by incoming air, and this seemed to work quite well, especially in Mexico and South Africa, which were obviously the most likely places for vaporization due to the combination of heat and altitude.

In scheming the layout of the DFV, I started off with lists of all the drives and accessories that would be needed. As we wanted to make a flat-faced engine, and to make it as short as possible, it was important to work out where to put these drives and all the accessories – things like the alternator, the distributor, the fuel injection metering unit, the mechanical fuel pump, the water pumps and the oil pumps. Finding homes for these, and providing drives for them, represents a very difficult scheme problem, and unless this is tackled as a one-man operation you tend to end up with something which looks like a Christmas tree. If one person deals with one item and somebody else with the next, you end up with a series of small fouls, and something which needs a complete redesign in order to accommodate one small part.

I did enough detailing to prove to my own satisfaction that I could get the oil pumps and scavenge pumps and so on down the side of the engine, and I drew sufficient of the metering unit, alternator housing and distributor to make sure that they would fit in the Vee. This had a certain amount of influence on the head design and the port configuration, because I had to leave more room than I would have liked in the Vee to provide space for the alternator housing. The actual detailing of the pumps and the metering unit and the alternator assembly was carried out by Mike Hall. We had a dividing line on the timing-case drive whereby I produced the drives for the front of the engine and dictated the line of the drives down each side. I was responsible for the cylinder block, the heads, the cams, the gear train at the front, the belt drive at the front and the rest of the engine, and I did the final details on the block and head drawings. The drawings for the crank, rods and pistons were either mine or a fair copy of them, and the same applied to all the covers on the engine; I would do a detail drawing but scribble in the dimensions rather than print them neatly, and then get a fair copy made.

I did all this work at home. I used to go to the factory one day a week to see about other problems and to discuss with Mike Costin and Mike Hall how the other bits were coming along. The rest of the time I would work from nine in the morning until midnight or later, and this went on for about nine months. I didn't find it particularly hard to do this, in fact I find it very easy to get interested in design, and at that stage I was sufficiently wrapped up to be quite happy in my work. As I was at home for all my meals I deliberately went on a diet, which seemed to be largely steak and cabbage, and as a result I lost over 40 pounds. But this was nothing to do with overwork or mental strain; I did suffer from eye-strain at one stage, but this was simply the result of doing too much drawing.

I used to think that I was very slow at designing, but when I think now of the amount I managed to get done on the DFV, and the amount of thought and detailing that went into it, I feel that my output was fairly high. Whenever there is a problem I try to think of all the possible ways of tackling it, and this often involves drawing several different schemes. The majority of designers seem to get the first idea into their head, carry straight on and detail it, and then have another thought as to whether the solution is getting unnecessarily complicated or whether perhaps there is an easier one.

Very few of the solutions which I arrive at strike me as good; generally it is a case of going for the one which is least bad. I am probably more pernickety than most other designers, but this is mainly because I try to get things right first time. Having stated that development is only necessary because of the ignorance of designers, I have a requirement to try and prove myself no more ignorant than is absolutely necessary. In order to design out the need for development it is necessary to think of all the problems that are liable to arise before committing yourself to metal. It is also necessary to simplify the manufacturing process by making sure that everything is dimensionally sound and will fit together, that all clearances and design margins are adequate and that the components involved will do the job. There seems to be a tendency for people to want to see a working model of their design to find out what is wrong with it; by looking at a drawing of the component I can at least say where it is likely to break and try to minimize the chances of its breaking.

Nowadays we have drawings of every component, and inter-stage machining drawings as well. We are also making drawings of components in their turned stage, or even in bi-operations, so that we can control their manufacture and make them the same in subsequent batches. To avoid having to decide how to make a component several times we produce written notes on each operation, drawings showing how bits will be held, and redimensioning drawings. These are intended for milling or turning operations, and the finished component can then be checked against an inspection drawing. We get a lot of extra drawings in this manner.

Some people have very complicated inter-stage drawings on every component. Others leave it to the operator to decide how something shall be made, but I feel that if you do this it is the operator who is designing the component. In some firms, of course, the operator's design could well be considerably better than the designer's!

By being interested in casting, machining and all aspects of engineering, I feel that I can bring to my design a fairly comprehensive knowledge of the whole problem, and therefore I tend to design simply. Having always found it very difficult to machine accurately, I attempt to provide limits which are as wide as possible. Should any design require limits that strike me as being difficult, then I attempt to redesign to avoid the difficulty; unless it is reasonably easy for things to be made it is unlikely that they will be made properly. There's no point in designing something that has a fair chance of working if you cannot control the way in which it is made. I would think this was the downfall of the Weslake V12, which as a conception was an exceptionally good engine. But it suffered from manufacturing problems, and from attempts to make it in an unsystematic fashion with inadequate control over accuracies and tolerances. It is necessary to have the machinery and the system to make things right – or near-enough right. Hacking around with files and scrapers isn't really on nowadays.'

5

1966: The return of power

WHILST WORKING on the FVA Formula 2 and DFV Formula 1 engines, Keith Duckworth was able to keep a distant watch on the Formula 1 scene, even though his busy schedule at Northampton prevented him from attending any world championship races and sizing-up the opposition at close quarters. Had he been able to do so he most probably would have obtained both enjoyment and satisfaction from his visits to the various circuits, for not only were several of the race results – by courtesy of Jack Brabham – helping to prove the wisdom of the simple V8 concept, but much of what he would have seen during his walks down the pit roads and through the paddocks would have convinced him that whatever the 12 and 16 cylinder engines might offer in the way of a challenge in the long term, there was little cause to fear them in their present state of development.

Jack Brabham had never been one of motor racing's great conversationalists, but he had always been one of the sport's deepest thinkers. Clearly, he had given the 3 litre formula a lot of thought, and in the circumstances he had achieved a very sensible solution to the short-term problems brought about by the change of regulations and the withdrawal of Coventry Climax.

Throughout his racing career Jack had thrived on driving relatively uncomplicated cars (his two world championships in 1959 and 1960 had been won with cars and engines which were models of simplicity in comparison with some of their contemporaries), and he saw no reason why this design policy should be altered by the introduction of a 3 litre engine limit. This was why he disfavoured the concept of a 12-cylinder engine, which would commit him to producing a large and relatively heavy car. He also declined the offer of the BRM H16 engine because despite its compactness it was no lightweight, and its mechanical complexity suggested that it would be a long time before it could be developed into a reliable race-winner.

Brabham, like Keith Duckworth, was convinced that a reliable V8 engine was the answer, and furthermore he knew the engine he wanted, at least as a stop-gap. For several years he had enjoyed a close association with the big Australian automobile component and equipment manufacturing concern, Repco Engineering (hence the Repco Brabham name on all his racing cars), and on one of his many visits to Repco's headquarters in Melbourne he had been shown their new racing engine project.

Even before Coventry Climax had pulled out of racing, Repco had been anxious to produce an engine which would replace the out-moded 2½ litre four-cylinder Climax which had been the mainstay of the annual Tasman Championship series of races in New Zealand and Australia. Repco had found a suitable cylinder block in the aluminium V8 which General Motors had developed for various Buick, Oldsmobile and Pontiac compact cars, but had since abandoned because of manufacturing costs. Repco's plan was to build special cylinder heads to mount on to the lightweight block, with a single overhead camshaft for each cylinder bank in place of the normal pushrods, and to build the engine in two sizes, a 2½ litre for Tasman races and a 4·2 litre for use in sports cars. Brabham realized that a 3 litre version was just what he wanted for his Formula 1 cars, and in his persuasive way he convinced Repco not only that they should build it, but also that they should give the Formula 1 version top priority.

The first 3 litre engine was installed in a chassis which Brabham's partner and chief designer Ron Tauranac had produced to accept the Coventry Climax flat-16 1½ litre engine, the chassis frame being modified at the rear to accommodate the V8. It was little larger than his 1965 car, and clearly it would be one of the lightest and certainly the most compact of the new grand prix cars.

After some satisfactory tests in England, Brabham shipped his new car to South Africa and came close to winning the non-championship Rand Grand Prix on the very first day of the new formula – January 1, 1966. The Brabham BT19 was the only full 3 litre car in the race, and Jack found his 300-odd b.h.p. more than sufficient to hold a comfortable lead until a few laps from the end, when the fuel distributor drive belt snapped following a failure of the rotor drive.

Nevertheless, Brabham was encouraged by the engine's performance and although he suffered another setback in the Syracuse Grand Prix – this time it was an internal water leak – he led the *Daily Express* International Trophy race at Silverstone from start to finish, and was only challenged seriously by John Surtees, who had been the race-winner in Sicily, in his new V12 Ferrari.

A week later, Brabham was feeling far from well when the world championship season opened with the Monaco Grand Prix, and he was not unduly upset when he was forced to give up with gearbox trouble. The next race was the Belgian Grand Prix at Spa, one which will be remembered for many years for the frightening series of accidents which occurred on the first lap as the tightly bunched pack went from a dry track into a wall of torrential rain just over a minute after the start. Car after car spun off the course, but Jackie Stewart, who was trapped in his 2 litre BRM for a time and suffered fuel burns in addition to his shoulder and rib injuries, was mercifully the only casualty from the multiple accidents. Brabham was one of the survivors, despite a lurid series of spins, and finding himself one of only seven drivers still in the race he decided that keeping his car in one piece was more important than a toss-up between maximum points and a write-off. He therefore drove steadily into fourth place and was thankful at the finish still to have a car in good shape for the French Grand Prix, not that he expected to make much of an impression on the ultra-fast Reims circuit which, given dry weather, was certain to favour the more powerful cars.

It *was* dry at Reims, and very hot, and as Brabham had anticipated the V12 Ferraris were easily the fastest cars. Nevertheless, Jack Brabham was to write a page of history that day as the first driver to win a world championship race in a car bearing his own name. Only Bandini was able to keep ahead of the Repco Brabham, which by this time was producing around 320 b.h.p., and when the leading Ferrari's throttle cable snapped the Australian inherited a comfortable lead over Parkes' Ferrari and was able to ease-up during the closing laps to score a very popular victory. Brabham's team-mate, Denny Hulme, who had been driving a Climax-engined car prior to this race, celebrated his first GP with the Repco V8 by finishing in third place, despite running short of fuel two laps from the end. There was great jubilation in the Brabham pit after the race, for not only had Jack realized a personal ambition, he was now leading the world championship table for the first time since his title in 1960 and had also scored his first world championship race victory in six years. Furthermore, the Repco Brabham's power deficiency had been proved of little consequence on a 'power' circuit, while the V8 engine's reliability had been proved without doubt. All in all there was a lot to celebrate that evening in the heart of the 'bubbly' country!

During the next few weeks it seemed that nothing could go wrong for Brabham. He led the British Grand Prix at Brands Hatch from start to finish, and with Hulme being the only other driver to complete the full race distance the Repco Brabhams scored their first 'one-two' victory on a circuit which was particularly suited to the cars' excellent handling characteristics and middle-range torque.

Brabham's third successive grand prix win occurred at Zandvoort, where Jack took time off for a spot of fun immediately before the race. Three months earlier he had reached the ripe old age (for a Formula 1 driver) of 40, since when many stories had appeared in newspapers and magazines referring to the 'grand old man' of motor racing, a journalistic trend to which Jack's diminutive but supremely loyal wife Betty had taken extreme exception. By this time Jack had also decided that there had been sufficient 'Too old at 40?' type stories, and he devised a way of putting an end to them.

While the rest of the drivers wandered along the track to their cars which were waiting on the dummy grid in full view of the main grandstand Jack waited behind at his pit then quietly slipped on a 'W. G. Grace'-type beard and whiskers. Then, when everyone started to wonder where he was, he grabbed a jack handle as a walking stick, and hobbled slowly towards his car which was awaiting him on pole position. The crowd loved it, although after this it was more important than ever that he should lead the race, and if possible win it.

At first everything went well, and Hulme was soon supporting Brabham in second place. But then Denny dropped back with ignition trouble and suddenly the circuit became incredibly slippery as a result of dropped oil. This helped to nullify the power advantage which Brabham held over Jim Clark's 2 litre Lotus-Climax, and with less than a third of the race run Clark had caught and passed Brabham. But Jack made a big effort as the race entered its final quarter, gradually closed the gap on the Lotus and retook the lead, his victory being assured a few laps later when Clark had to stop for more water. The 'old man' had won and proved his point.

If Brabham had been given a tough race in the Dutch Grand Prix his German Grand Prix victory was even harder-earned two weeks later on a Nurburgring made treacherous by intermittent rain. In this case it was that *Ringmeister* Surtees who provided the main challenge, driving a Cooper-Maserati with which he led the race initially then clung to the tail of the Repco Brabham until the loss of his clutch and most of his gears forced him to drop back during the last two laps.

This was to be Brabham's last GP win of the year, but he had already collected sufficient points to make him almost certainly a world champion for the third time, and his title was confirmed in the Italian Grand Prix while he was sitting on his pit counter at Monza, having retired after a few laps with a major oil leak. Only Surtees had had a mathematical chance of overtaking Brabham's total, and this disappeared a few minutes after Jack's retirement when the Cooper-Maserati developed a fuel leak. Hulme put in a dogged performance with the surviving Repco Brabham, and remained in close contact with the Ferraris of Scarfiotti and Parkes right to the end to finish a gallant third and put his team closer to the constructors' championship.

The second title was to be settled during the United States Grand Prix, for although both Repco Brabhams retired, Hulme's with lack of oil pressure and Brabham's with a broken cam follower, the lone Ferrari – the only remaining challenger – also dropped out with engine trouble. The Brabham team, therefore, went to the final round, the Mexican Grand Prix, as double world champions, and although they were unable to match the speed of Surtees' Cooper-Maserati on this occasion they did the next best thing by finishing second and third, Hulme being a lap behind the boss after spinning off and having to restart with the car jammed in fourth gear. It had been a remarkably successful year for the Brabham Racing Organization, with achievements far beyond Jack's most optimistic hopes. He would have been disappointed had his cars not scored at least two victories during the 1966 season on the strength of their anticipated reliability, but it had come as a most pleasant surprise to him that the opposition had not been stronger, and that he was able to score three of his four championship race victories through speed as well as stamina, only the French Grand Prix having been won as a result of a faster car running into trouble.

The only other British works team to run the full season with 3 litre cars were Cooper, who had built a strong-looking but inevitably rather large and heavy car around the equally bulky Maserati V12 engine, a power unit with an ancestry dating back to the 2½ litre grand prix formula of the late Fifties, and which had been seen from time to time in the intervening years in larger-capacity forms as a sports car engine. One advantage which Cooper derived from their association with Maserati was their ability to offer a complete grand prix car for sale to customers, for the Italian company were able to offer a regular supply of V12 engines. As a consequence, there were several customers for the Cooper-Maserati during the early months of 1966, and Jo Siffert, Jo Bonnier and Guy Ligier were seen regularly in them in support of the works cars driven by Jochen Rindt, Richie Ginther (pending the completion of the 3 litre Honda) and later in the year John Surtees, following his break with Ferrari.

Like the Brabham team, Coopers, who had become part of Jonathan Sieff's Chipstead Motors Group a few months earlier and had moved from their traditional home at Surbiton to new premises at Byfleet, found themselves with a spare space-frame chassis when the 16-cylinder Climax engine had failed to materialize, and like Brabham they converted it into their prototype 3 litre Formula 1 car. However, they intended their 'production' car to be a monocoque, which seemed a much more sensible form of construction for a car which would require a large fuel capacity to satisfy the considerable thirst of the big Maserati engine.

The first Cooper monocoque was unveiled at the Racing Car

Show in London in January 1966, and it made its first race appearance at Syracuse three months later, when ignition trouble robbed both Siffert and Ligier of any chance of success. Also, bent front wishbones suggested that even at 120 pounds or more over the minimum weight limit the cars still required some beefing up. Many hundreds of test miles were carried out in England, but when the works cars made their first British race appearance in the International Trophy at Silverstone they were still clearly in need of further chassis development, while engine power was obviously still insufficient to offset the burden of the cars' high startline weight.

That the cars were still uncompetitive at Monaco was no surprise, for there could hardly have been a less suitable circuit for a bulky car with a modest power-to-weight ratio, but morale was restored considerably at the Belgian Grand Prix where Jochen Rindt set the second fastest practice time and went on to finish in second place having led Surtees' winning Ferrari on no less than 19 of the 28 laps before his differential packed up. There was a strong element of bravery in this performance, for the circuit was treacherous throughout the race, but nevertheless it was a promising result for Cooper, with Ginther and Ligier finishing fifth and sixth, and the only retirements (by Siffert and Bonnier) being the result of first-lap accidents.

Surtees' victory was followed a few days later by his big row with Ferrari, as a result of which he left the team, and through the generosity of Shell, to whom he was under contract, he appeared for the French Grand Prix as a member of the BP-backed Cooper team in place of Richie Ginther. Surtees derived considerable enjoyment in setting a practice time at Reims bettered only by Bandini's Ferrari, from which he had obtained a valuable tow, but he was soon out of the race with injection pump trouble, and it was left to Rindt to lead the Cooper effort, although there was little he could do on that very hot day with an engine which was intermittently overheating and suffering fuel starvation.

Modified fuel systems were tried on the Cooper-Maseratis in time for the British Grand Prix, and although these were a considerable improvement, Brands Hatch exposed the cars' handling deficiencies, and both Surtees and Rindt, who were running well during the early part of the race, gradually dropped back as their cars' roadholding deteriorated, Surtees eventually retiring with virtually no shock absorbers and Rindt finishing fifth.

In Holland, Rindt crashed his car on the second lap after finding a wrong gear, and Surtees, who had had to take over a spare car after his own had blown its engine in practice, had a miserable race before retiring with a dead battery, so once again Cooper were going through a bad spell.

However, the season's best result so far came from the German Grand Prix by which time a little more power had been found from the Maserati engine and some of the chassis deficiencies had been rectified. The team was already beginning to benefit from the arrival of Surtees on the scene, for although John was invariably outspoken in his comments and intolerant of anything other than a 100 per cent effort from everyone concerned, he usually talked good sense when it came to improving the performance of the cars, and it was no coincidence that some real progress was beginning to be made in this direction.

Surtees, of course, invariably has driven well at the Nurburgring, and he put in a great effort for the 1966 German Grand Prix, claiming second-best practice time, keeping Jack Brabham close company nearly all the way to the chequered flag, and setting the fastest race lap on a damp and very slippery circuit. With Jochen Rindt following him home in third place, the race was a considerable tonic for the Cooper team.

After that performance, Surtees' fuel leak in the Italian Grand Prix the following month was a considerable blow, because his car had been going very well indeed in the leading bunch, the engine seeming to have responded successfully to an experimental conversion to coil ignition. However, with fuel spraying on to a rear tyre the car was almost uncontrollable on right-hand bends and he was forced to give up. Jochen Rindt had the cruel misfortune to suffer a puncture with just over a lap to go, but he was sufficiently far ahead of the next car to maintain his fourth place to finish, where he passed the flag on three tyres and a wheel rim.

The United States Grand Prix result looked healthy enough on paper for the Cooper team with Rindt, Surtees and Siffert (in Rob Walker's car) taking second, third and fourth places behind Clark's Lotus-BRM, but this was a race which the Cooper-Maserati could well have won. Surtees was about to lap Arundell's

Lotus-Climax quite early in the race when the two cars touched and spun off course, the ensuing pit stop to inspect the Cooper-Maserati for possible damage causing further valuable loss of time. Once he had been assured that all was well, Surtees went off like a rocket, fought his way back into third place, and set a new lap record for the Watkins Glen circuit. But of course he had lost too much ground.

But compensation came in the final race of the season at Mexico City where, despite a violent bout of 'Montezuma's Revenge', one of the perils of the GP finale every year, Surtees gave Cooper their long-hoped-for first victory by a margin of just under eight seconds ahead of Jack Brabham. Although both Rindt and Siffert dropped out with suspension failures after losing balance weights from their front wheels, Surtees' car ran faultlessly, its engine seeming to be well tuned to the rarefied atmosphere. Also, the race did more for John's illness in two hours than pills and potions had managed in two days, for in the early laps he perspired so much that most of the poison left his system, with the result that he felt in fine shape by the end of the 65th and final lap. Of course, being in first place must have helped!

Surtees had been lying on his back in St. Thomas' Hospital in London when the 3 litre formula had been introduced the previous January – the legacy of a nasty accident with his Lola T70 sports car in Canada. Nevertheless, Ferrari, for whom he was then number-one driver, were still considered to be the team most likely to succeed in the first year of the new formula.

By the time 'Big John' was mobile again there were two cars ready for him to test, a brand-new design with a spaceframe chassis reinforced by stressed-skin panels and powered by a new twin-camshaft version of the 3 litre Ferrari V12 single-cam sports car engine, and a lightweight car which had been prepared for the Tasman Championship (but not used), and which had been fitted with a 2·4 litre version of the Dino V6 engine.

Surtees became the first driver to win a Formula 1 race with a full 3 litre car when he drove to a comfortable victory in the Syracuse Grand Prix, with Bandini backing him up in second place with the V6. At that stage the V12 was only producing about 330 b.h.p., against the 260 b.h.p. of the V6, but the ease of Surtees' victory suggested that the rival Maserati V12 was probably delivering little more than 310 b.h.p. on the day, and that talk of the new 3 litre engines giving anything from 360 b.h.p. upwards was, to say the least, premature. Encouraged by their initial success, Ferrari sent the V12 to Silverstone for Surtees to drive in the International Trophy race, and although they were beaten narrowly by the new Brabham they gained some valuable test experience during the race; this was the first time the car had been driven on a fast circuit in the dry, and the roadholding, which had seemed satisfactory at Syracuse, was not yet a match for the Brabham through the sweeping curves of Silverstone. Nevertheless, the Ferrari was full of promise.

However, the first rumblings of discontent between Surtees and his employers was to be seen at Monaco a week later when he was told that he would be driving the V12, when clearly the V6, which which was handed to Bandini, was the better bet on this slow and twisty circuit. After a big practice effort he managed to put the 3 litre car on to the front row of the grid, and then led the race for 14 laps until his differential failed, but from then on it was left to Bandini to chase Stewart's leading 2 litre BRM. The Italian put in a very spirited drive, breaking the lap record, but he wore out his brakes in the process and after closing to within 11 seconds of Stewart he was obliged to drop back during the last few laps to finish a lonely second.

The Belgian Grand Prix was to be Surtees' last race for Ferrari, before his big row and walkout at Le Mans a few days later, but he left them with a worthy win after a finely judged race in terrible conditions against Rindt's Cooper-Maserati during which he lapped Bandini's third-place V6-engined car.

Mike Parkes, who until then had been mixing his duties as a Ferrari development engineer with some sports car racing, but had been hankering after a Formula 1 drive for a long time, finally saw his chance on Surtees' departure, and a special long-wheelbase car was built to accommodate his lanky frame in time for the French Grand Prix. This was the first time that Ferrari had run two 3-litre-engined cars, and for more than half the race it looked as though Bandini would celebrate his promotion by scoring a runaway win in the ex-Surtees car. By this time the V12 engine was producing over 350 b.h.p., and the Italian was able to dictate the pace with ease, having already proved his car to be easily the fastest in practice. But so often the 'easy' races provide the biggest disappointments, and Bandini freewheeled to a stop with a broken throttle cable just when victory seemed to be in the bag, while Parkes went on to finish second to Brabham in his first race with his new car.

A wave of strikes in Italy prevented the Ferraris from appearing at Brands Hatch and so a renewal of the Ferrari-Brabham battle had to wait until the Dutch Grand Prix, by which time the Italian cars had lost some of their competitiveness. Bandini crashed his car in practice and drove a subdued race into sixth place, while Parkes missed a gear early in the race and ran off course into Rindt's abandoned Cooper, badly damaging both cars.

But despite these setbacks Ferrari ran a three-car team at the Nurburgring, Scarfiotti joining in with the V6-engined car with which he surprised a lot of people, not least of all Bandini and Parkes, by being the fastest Ferrari driver in practice and earning a place on the front row of the grid. But Scarfiotti dropped out with ignition trouble, Parkes left the circuit after his throttle jammed open, and Bandini found his car so uncompetitive in the wet that he was lapped by Brabham before finishing in sixth place.

The Italian Grand Prix, of course, is the most important race of all for the Ferrari team, and at Monza that year there were three V12-engined cars with new three-valve cylinder heads plus the V6, which on this occasion was handed to Tim Parnell to run for Baghetti. But the team's main hope, Bandini, was in trouble right at the start of the race, and eventually dropped out with ignition failure, while Baghetti lost a lot of time with a pit stop after running off-course. However, everything worked perfectly for Scarfiotti and Parkes, who worked well as a team (supported by Bandini between his pit stops), Parkes managing to keep Hulme at bay while Scarfiotti made a break to gain a highly popular home victory.

This was the high point of Ferrari's grand prix year because they decided to send only one car to the United States Grand Prix, for Bandini, and when a plug electrode dropped into the engine and wrecked it the team decided to forget about Mexico and return to Europe to prepare for the 1967 season. They had not achieved as much as some people had anticipated in their first year of 3 litre racing, but neither had they been disgraced. The 2·4 litre V6 had proved to be a useful car on some circuits, but clearly the future had to be with the V12, which had shown itself to be fairly reliable and powerful enough to deal with most of the opposition which had come its way. If the car had any shortcomings they seemed to be in the chassis department, for handling, though good on some circuits, had been inferior on others, and the Ferrari was certainly not at its best in the wet.

One of the big disappointments of the year was BRM's H16 engine, which potentially had been the most powerful 3 litre of all, but with one notable exception had proved hopelessly unreliable. Also, during its first year it was developing nothing like the 400-odd horsepower which had been anticipated.

BRM had first shown their new car and engine during practice at Monaco, and although there had never been any intention to use them in the race (the ex-Tasman 2 litre V8s were far more suitable for this circuit) it was expected that the 3 litre cars would spear-head the BRM attack from the Belgian Grand Prix onwards. However, both car and chassis needed a great deal of development work, not to mention a certain amount of redesign, and it was not until the Italian Grand Prix, following an edict from the management, that the cars were finally used in a race. Even then they disappointed, for Hill's car was out with a broken engine a few seconds after the start and Stewart's retired with a fuel leak just when it was beginning to show some promise. At Watkins Glen both cars had retired before half-distance, Hill's with a broken crownwheel and pinion and Stewart's with a broken engine, and in Mexico both engines failed comparatively early in the race.

It had been left to Team Lotus first to show sufficient faith in the H16 engine to use it in a race, and later to demonstrate that it could win. Lotus had decided to use the BRM engine as a stop-gap pending the arrival of their Ford DFV V8s in 1967, and had built three chassis which, like the works BRMs, had short monocoque chassis terminating at the rear of the cockpit, with the engine bolted direct to the back bulkhead. Lotus had intended running one of the cars for Arundell at Spa, but a practice engine failure prevented this. However, Arundell did start a Lotus-BRM at Reims, although it lasted only three laps before losing its gears.

Jim Clark's first race in the 3 litre Lotus-BRM was in the Italian

Grand Prix where, despite several pit stops, he was able to demonstrate the car's potential by almost equalling Scarfiotti's new lap record and keep going until two laps from the end when once again the car ran short of gears. Encouraged by this performance, Clark and Chapman decided to risk the H16 in the United States Grand Prix and were rewarded by their only victory of the year. Having retired both their own cars during the race, BRM didn't know whether to rejoice or weep at this surprise success because Clark's victory had been won with an engine they had lent Lotus on the eve of the race, after their own had broken! But Clark's victory – and it had been a very close-run thing because the engine was very sick by the end of the race – was only a temporary bright spot in the H16's otherwise fruitless year, and Jimmy became the first retirement from the Mexican Grand Prix, once again a victim of gear-selection problems.

Like BRM, Lotus relied principally during 1966 on their previous year's cars with uprated 2 litre engines, but whereas Stewart had found his Tasman BRM a race-winner at Monaco, Clark and Arundell were out of luck with their Climax and BRM-engined Lotus 33s, the closest thing to a victory being Clark's great drive in the oily Dutch Grand Prix before his water pump failed him in the closing laps.

The lack of suitable 3 litre engines meant that Bruce McLaren could scarcely have chosen a less opportune moment to branch out into grand prix racing, although he made a brave effort to put a car on the line for the first world championship race of the new formula. With designer Robin Herd he produced a neat monocoque chassis, part of which was fabricated in Mallite 'sandwich' material consisting of balsa wood compressed between two skins of aluminium. At that time McLaren was quite heavily involved with Ford in the United States in the endurance racing field, and he had been attracted by the 4·2 litre Ford four-overhead-camshaft V8 engine which had been doing so well at Indianapolis. His idea was to destroke this engine to conform to the 3 litre Formula 1 limit, and use it as a temporary expedient (it was too heavy to consider as a permanent choice) in his first GP car. Unfortunately the engine lost its oil during the Monaco race and was quite badly damaged, so McLaren was offered a substitute V8 from the Serenissima concern in Italy for subsequent races. Although this engine was no more powerful than the converted Ford (they were both delivering about 300 b.h.p) it was considerably lighter, having been designed for a relatively small sports car, but it was under-developed and Bruce was unable to bring it to the starting line until the British Grand Prix, in which he managed to finish in sixth place. More practice blow-ups followed as the season continued, and the Formula 1 McLaren was not to race again until the last two grands prix, by which time the Ford engine had been reinstalled. Bruce's best result in a lean year came at Watkins Glen with a fifth place, but he had to give up with engine trouble two-thirds of the way through the final race at Mexico. By then he was beginning to think that perhaps sports cars were a lot more fun than Formula 1 cars!

The anticipated new cars and engines from Honda and Dan Gurney's Anglo American Racers team helped to enliven the later grands prix, although Gurney's prototype Eagle, fitted with a 2·7 litre Climax engine, had appeared as early as the Belgian Grand Prix. It was during this race that the tall Californian caused a big laugh by parking his car at the trackside, leaving the engine running like a taxi, rushing off into the bushes, and returning a minute or two later – looking very relieved – to carry on motor racing!

The Eagle, designed by Len Terry, seemed to have particularly good road-holding, and after scoring a fifth place in the French Grand Prix Gurney surprised his competitors by claiming a place on the front row of the starting grid at Brands Hatch, although on this occasion his engine let him down. He also retired at Zandvoort and finished without any championship points at the Nurburgring after dropping from fourth place with an electrical problem. But the Eagle seemed to be a car with a future, and the first of the new chassis with the Westlake V12 engine was a centre of interest at Monza, for Harry Weslake was a known authority on gas flow problems and his first grand prix engine was expected to deliver a lot of power. It was certainly a superb looking piece of machinery, sitting squatly in the Eagle chassis with four huge megaphone-type exhausts extending out behind the immaculately finished cam covers. But it was still short of test hours, and Gurney retired from the race after only seven laps with a fuel leak and an ominously high oil temperature. He also retired early from the United States Grand Prix, while Bob Bondurant, who had been given the

Climax-engined car, also had to drop out after damaging his suspension in an incident with a 2 litre Lotus-BRM driven by Rodriguez. The Weslake engine was still very fragile, though producing well over 380 b.h.p. when running properly, and so Gurney exchanged cars with Bondurant for the final race, which proved a wise move because he finished in fifth place whereas once again the V12 Eagle was put out with engine trouble.

The Honda, like the Eagle-Weslake, was first seen in action at Monza. This time the Japanese had seen fit to mount their V12 engine longitudinally in a relatively conventional chassis (their 1½ litre V12 had featured a transverse engine). The new car appeared to be something of a heavyweight, however, which meant that it was by no means overpowered with its claimed 370 b.h.p. However, it seemed to have very good acceleration, and no doubt Richie Ginther was thankful for the car's substantial construction when, after moving up to second place in the early laps of the Italian Grand Prix, he punctured a rear tyre and went off-course at about 140 m.p.h., fortunately emerging with nothing worse than a damaged shoulder.

The Honda was a write-off, but the team had two replacements in time for the United States Grand Prix as evidence of their serious intent under the new formula, Ginther being rejoined by Ronnie Bucknum. However, both engines seemed to be in trouble with their fuel mixture, and Bucknum retired with a serious shortage of power, while Ginther struggled on to the finish, although too far behind to be classified. Returning to the scene of his success the previous year, Ginther managed the third fastest practice time in Mexico, but although he streaked into the lead at the start, the Honda started to drop back as it ran into fuel vaporization problems. Nevertheless it ran consistently and reliably at its reduced speed to finish fourth, and there was a new lap record in the bag as well to bring smiles back to the faces of the inscrutable Japanese in the pits. Bucknum also completed the race, but not without incident because at one stage his battery short-circuited and set light to his seat. There have been many causes for pit stops during the history of grand prix racing, but surely Bucknum added a touch of history by pulling in with his seat on fire!

So ended the first season of 3 litre Formula 1 racing. If nothing else the change in regulations had brought about a level of variety not seen in grands prix for many years. There had been V12 engines from Ferrari, Maserati, Weslake and Honda and an H16 from BRM. As a temporary expedient, Formula 1 engines had been produced out of Ford and Serenissima V8s which had been designed for other purposes, while BRM and Lotus had prepared backup cars with 2 litre V8 engines which in the event were to form the mainstay of their racing efforts. AAR Eagle were also grateful for the reliability of an obsolete, under-sized engine for several races, while Ferrari were very well served early in the year by their 2·4 litre V6. Completely new chassis were produced by BRM, Cooper, Ferrari, Lotus, Honda, Eagle and McLaren to be supplemented in several cases by designs carried forward from the previous formula.

It was, therefore, a year of great technical interest, if only for the fact that when the time came to add up the score, every one of these cars and engines had been beaten on points by Jack Brabham's team of Repco Brabhams. Both the drivers' and constructors' world championships had been won with a car with such 'out-of-date' features as a spaceframe chassis and outboard-mounted front suspension units, and powered by a simple V8 engine which had been developed from the cylinder block of a production car, and at the beginning of the year had been delivering no more than 300 b.h.p. and at the end of it no more than 330.

Admittedly the opposition had not been at full strength, but then few people had expected it to be so early in the life of the formula, least of all Jack Brabham. As recently as mid-1965 he had been insisting that there would be no way in which he could afford to become involved in the costly power race which the new regulations would create. But even while he had been making his 'count me out' noises his plans had been formulated. Maybe it was his good fortune that he had a Repco connection, but it had been his shrewd evaluation of what was needed during the first years of the 3 litre formula which had provided the foundation for his success. He had listened to all the arguments in favour of power and complexity, had carefully avoided challenging them in open debate, and then had slipped away quietly to do the job his own way. It was typical Brabham strategy, and as usual it had worked!

6

1967: V8 for Victory

THE MAY 1967 deadline for the completion and delivery of the first Ford Formula 1 engine had been fixed in the optimistic hope that the Lotus 49 would make its public debut in the Monaco Grand Prix on May 7. It meant a desperately tight schedule for Cosworth Engineering, but they managed it with a few days to spare, the first engine being handed over to Lotus on April 25.

Meanwhile, the new Lotus had been taking shape behind closely guarded doors at Hethel, Cosworth having supplied a mock-up engine to enable Colin Chapman, his chief designer Maurice Phillippe and their staff to finalize the oil and water plumbing, the electrical connections, and the installation of all the ancillary equipment. But even with all this work complete there was still a lot to be done when the first working engine arrived at the Team Lotus headquarters.

First, there was the critical moment when car and engine were brought together, and every nut, bolt and connection had to be checked and double-checked before the engine could be fired up for the first time. This is always a tense and exciting operation, and it tends to be performed to a set ritual as everyone involved gathers around the car, peering intently at the gauges, and looking for a tell-tale sign which could mean that the cooling or lubrication systems are not functioning correctly. During this period you can be sure that someone will be standing within easy reach of the 'kill' switch, ready to cut off the electrics in a flash should something go wrong. But if all is well, with the oil pressure, oil temperature, water temperature and fuel pressure gauges all showing healthy readings as the engine warms up, and every union proving to be leak-free, the first big hurdle will have been passed, and the serious business of mobile testing can be started.

Test driving unquestionably is the most dangerous part of a racing driver's life, particularly when he has to evaluate a brand-new, untried car. It is also the moment of truth for the designers, who will have checked, double-checked, and checked yet again their calculations, convinced themselves that they have produced the best possible job, and seen their design go together like a well-cut jigsaw puzzle, yet who know that the only real test is how it will perform in the hands of its driver. Will it be an 'easy' car to drive (the word is used in a relative sense because no racing car is ever easy to drive to its limit), or will it prove vicious, perhaps dangerously so? If it feels 'easy', is this because it is not fast enough, and conversely, if it feels 'twitchy', is this because its cornering performance is higher than anything encountered before and therefore taking the driver into unfamiliar territory?

The solution to these and a hundred-and-one other questions aimed at evaluating and if possible improving the performance and efficiency of the car, and making life as comfortable and as safe as possible for the driver, is vital to the success of any racing team, and it is the constant though rarely fulfilled aim of entrants to have much of this work completed before the start of a new racing season.

In the case of the Lotus 49, of course, this was impossible, but nevertheless Colin Chapman, Keith Duckworth and Walter Hayes were unanimous in their view that Graham Hill, who was to be responsible for most of the development driving of the Lotus 49 due to Jim Clark's absence abroad, should be given an opportunity to test the new car thoroughly before it was committed to a race.

It was agreed, therefore, that the idea of a Monaco debut should be abandoned, and that two Lotus-Fords would be made ready for the next round in the world championship series, the Dutch Grand Prix, which was scheduled for June 4 at Zandvoort. Those extra four weeks were to see some hectic activity both at the Cosworth Engineering factory at Northampton and at the Team Lotus headquarters and adjacent test track in Norfolk, although the test programme was to be hampered considerably by the fact that for much of May Graham Hill was engaged in testing, qualifying and racing at Indianapolis.

Meanwhile, the second season under the 3 litre formula had started in South Africa on the same Kyalami circuit which a year and a day earlier had staged the non-championship Rand Grand Prix. Mike Spence had been the victor in 1966 with a 2 litre Lotus-Climax after Jack Brabham's new 3 litre Repco Brabham had retired from the lead, and now, with a well-developed car at his disposal and the world champion's crown on his head Jack was in a determined mood to avenge his earlier misfortune.

This time the race carried the title of the South African Grand Prix, and as such it became the first round of the 11-race fight for the 1967 title. Consequently, the organizers had been able to attract a high-quality field, although Ferrari decided to stay at home to carry out a test programme on their V12-engined car, and Dan Gurney elected to run just his Climax-engined Eagle while Harry Weslake's people continued to build more reliability into their own 3 litre engine. The main driver change had been Graham Hill's return to Lotus after spending the previous seven years with BRM, Graham's place at Bourne having been taken over by Jackie Stewart, who now had Mike Spence as his number two driver. John Surtees had ended his short stay with the Cooper-Maserati team and had accepted an invitation to join Honda who, although they were to run only one car, at least for the time being, had promised Surtees their full support. This left Richie Ginther without a drive, although his fellow Californian Dan Gurney was quick to offer him the seat of the second AAR Eagle, which Ginther was happy to accept, even though it meant that he would have no car to drive in the South African Grand Prix.

Jack Brabham was easily the fastest in practice at Kyalami, and with Denny Hulme alongside him on the front row with the second Repco Brabham BT20 it looked as though the 1966 champion car and driver would still prove a formidable combination. Jim Clark managed to work his H16 BRM-engined Lotus 43 on to the second row alongside Pedro Rodriguez in the works Cooper-Maserati, and John Love of Rhodesia made some large local headlines by claiming a third-row position for his 2·7 litre four-cylinder Cooper-Climax, a former Tasman Championship car, alongside the Surtees Honda and ahead of Jochen Rindt's Cooper-Maserati.

At first all went well for the Brabham team, Hulme making the best start (which was quite a rarity for Denny, who is inclined to take his car off the line fairly gently, then work hard to gain ground after the race has settled down). Jack Brabham immediately tucked in behind his team-mate and might have stayed there had he not touched cars with the Honda and spun wildly down the track. However, he recovered quickly, and soon after quarter-distance he had repassed the cars which had overtaken him during his spin, and was back in second place again behind Hulme.

But there was trouble ahead for the Brabham team. First Jack's engine cut dead and he limped slowly to his pit, where he spent a long time before eventually restarting with the engine still misfiring badly, and struggled on to finish sixth. Hulme's trouble began just as he was about to enter the final quarter of the race. Although he was holding a comfortable lead he was having to pump his brakes every time as they were losing fluid. With the brakes almost gone he rushed into his pit, told his crew the trouble, then rushed out again while they found some more fluid. The next lap he came in again, the reservoir was topped up, and Hulme shot off again, determined to re-take the lead from Love's venerable Cooper-Climax. But Denny was still in brake trouble and he had to make a third stop to have the hydraulic system bled, which dropped him back to fourth place.

It looked as though an 'undersized' car was again to win a Formula 1 race on the Kyalami circuit, for Pedro Rodriguez, who was in second place with his Cooper-Maserati after Rindt's car had dropped out with a broken engine, was hampered by the lack of second gear, while Surtees, lying third with the Honda, was dropping back with an overheating engine and deteriorating tyres.

Love's Tasman Cooper had a smaller fuel capacity than was necessary for a 203-miles grand prix, and so his mechanics had rigged up a supplementary tank with a separate pump to see him through. But unfortunately the pump was not operating properly, and a few laps from the end Love's engine started to splutter and he tore into his pit to the groans of the undertandably partisan crowd lining the circuit. Two gallons were poured into the main tank very quickly but by the time Love could rejoin the race Rodriguez was past and into such a secure lead that team manager Roy Salvadori was able to hang out a 'Take it easy' signal to the surprised but delighted Mexican, who went on to give Coopers their second successive win in a Formula 1 race. But there might have been less jubilation in the Cooper team that evening had they known that it was also to be their last.

Coming so soon after the final race of the 1966 season, the South African Grand Prix offered no reliable guide to the strength of the opposition which the Ford DFV engine might meet when it appeared in a few months' time, particularly as two of the potentially strongest engines, the Ferrari and Weslake V12s, were absentees. Certainly nothing which happened at Kyalami suggested that the BRM H16 was likely to be a serious threat for quite some time; both Stewart and Spence had dropped out early with engine failure, Hill's Lotus 43 had left the track and damaged its suspension after a few laps, and Clark's car overheated badly and retired just after quarter-distance. However, the next round in the world championship, the Monaco Grand Prix, was a full four months away, which meant that everybody now had time for some serious development testing, with the opportunity to try out their cars and engines in a trio of non-championship Formula 1 races to be held in Britain during March and April.

The first of these was the *Daily Mail* Race of Champions, held at Brands Hatch in the form of two 10-lap heats and a 40-lap final, with every car entitled to take part in both heats. It developed into a magnificent series of races, with the Eagles of Gurney and Ginther locked in combat with Surtees' Honda in the two heats, Gurney winning both with Surtees finishing second ahead of Ginther in the first and Ginther reversing the order in the second. Then Gurney went on to lead the final all the way, staving off a tremendous four-car challenge to receive his third chequered flag of the day, with a bare 4·2 seconds covering the first five cars. Dan's win, the first for the Eagle-Weslake, was highly popular, extremely exciting, and as it turned out subsequently, exceedingly fortunate, for as the Eagle crossed the line it was almost out of oil.

All American Racers had set up shop behind Harry Weslake's engineering business at Rye, and both companies had been a hive of activity during the winter months. While Eagle-designer Len Terry was busying himself with a series of chassis modifications aimed at reducing the car's weight from the 1,322 pounds it had scaled at Watkins Glen to around 1,250 pounds, Michael Daniel,

Weslake's son-in-law and technical director, was overseeing the development programme on the V12 engine.

A series of test sessions at the disused Goodwood race track brought some very encouraging lap times, and no signs of the crank-case pressurization problems which had dogged the engine on its early appearances, so it looked as though some real progress had been made. Team manager Bill Dunne, a former motoring writer, photographer and club racer in California, and an old chum of Gurney's, was confidently predicting around 400 b.h.p. in time for the Monaco Grand Prix, having already seen nearly 390 b.h.p. at 9,250 r.p.m. on the Weslake test bed.

However, the Race of Champions, though providing a magnificent morale-booster for the team, also revealed that there were still bugs to be removed from the Eagle's engine. The excessive oil consumption, which suggested a problem with piston rings, was not confined to Gurney's car, which had been trailing smoke for much of the final; Ginther's car was also found to be low on oil when Richie dropped out from second place three laps from the end with a steering problem. Nevertheless, the Weslake engine clearly had a lot of power, and this had enabled Gurney to build up such a useful lead in the final that he was able to ease up as he saw his oil pressure drop, and just hold off Bandini's 36-valve Ferrari, Siffert's Cooper-Maserati, Rodriguez's works Cooper-Maserati and Scarfiotti's Ferrari to the flag.

The result might have been different, however, had not Denny Hulme's Brabham broken a timing chain in the second heat and Jack's car lost an ignition lead just when he was closing in on Gurney in the final, for there was no doubt that the Brabham team were still a major power in Formula 1 racing, despite their modest 330 horsepower. They were able to prove this most convincingly a few weeks later at the International Spring Cup meeting at Oulton Park, a *Daily Express*-sponsored charity event in which the teams appeared without starting or prize money, the proceeds of the meeting being presented to the recently established International Grand Prix Medical Service and other voluntary medical organizations.

Unfortunately, there was only a small turn-out of cars and drivers, and with the Eagles posted missing presumed dismantled it was left to John Surtees to provide the main opposition to the two Repco Brabhams with his Honda. Jack Brabham was particularly anxious to take part in this race in order to try out the interim stage of his 1967 engine modifications. In order to extract more power from the Repco V8 it had been decided to build new cylinder heads, with the exhausts emerging from the centre of the engine instead of from the sides, and to mount them on a new Repco-made cylinder block. They were hoping that this would give them around 350 b.h.p., but as a temporary measure the new cylinder heads were mounted on one of the 1966 GM blocks for Jack to use in his car at Oulton Park as the first of the new blocks was not expected to arrive from Australia until May or June.

The new heads gave Brabham about 10 b.h.p. more than Hulme, but it was Denny who led the first 10-lap heat from start to finish to record his first victory in a Formula 1 race. Jackie Stewart had managed to keep his BRM firing on all 16 cylinders to set the fastest practice time, but he dropped out with a broken fuel pump drive on the third lap, while Jack Brabham's challenge from second place ended on the eighth lap when his ignition failed. After that Hulme had an easy drive to the finish ahead of Surtees' Honda and Spence's BRM.

The second heat brought another Hulme victory ahead of Surtees, while Brabham worked his way up from the back of the grid to finish third ahead of Stewart's repaired BRM, and the final developed into a Brabham demonstration run. This time Jack beat Denny into the first corner, and the two of them completely dominated the race, gradually pulling away from the Honda and building up a lead of over 20 seconds before crossing the line nose-to-tail. Surtees had been spared the sticking throttle which had caused his retirement from the final at Brands Hatch and had put up a dogged performance in the Honda, but Oulton Park is an uncharitable circuit to an overweight car, and the Repco Brabham's grip on this race was never seriously threatened. As no substantial Honda weight-saving seemed possible without a major chassis redesign it looked as though Surtees had signed up for a season's muscle-building!

The final non-championship Formula 1 race before the Monaco Grand Prix was the traditional *Daily Express* International Trophy at Silverstone, where the lure of both starting money and prize money attracted an entry only slightly larger than the modest

Oulton Park turn-out, the risk of a major blow-up or an accident only eight days before the next world championship round being a major deterrent to most teams. The Eagles and the works Cooper-Maseratis were missing, and only single-car entries were sent by BRM, Ferrari and Lotus to challenge the two Brabham BT20s, both of which now had engines equipped with the centre-exhaust cylinder heads. As Hill only had a Lotus 33 with a 2 litre BRM engine it looked as though the 150-miles International Trophy would be just a four-car race, with possibly Jo Siffert's Cooper-Maserati entered by Rob Walker making a fifth contender.

In the event it became a one-car demonstration, for Mike Parkes, driving his 1966 long-wheelbase Ferrari with the 1967-type V12 engine with three valves per cylinder, led from start to finish, and apart from an early challenge by Stewart before the BRM broke a universal joint he was completely unmolested. The only excitement developed in the fight for second place, for Brabham found himself busily engaged with Siffert after sliding on oil early in the race and losing ground, and the two exchanged places several times before Brabham finally nosed in front of the Cooper-Maserati on the last lap, a manoeuvre which so upset Siffert that he spun his car on the slowing-down lap! Denny Hulme had been an early retirement after striking a fresh patch of oil and sliding backwards off-course, and Graham Hill finished a lap behind in third place after a pit stop to change an oiled-up plug. Not the most memorable of races, therefore, but one which served as a reminder that Ferrari, even when running only one car, should never be underrated.

One week later further evidence of Ferrari potential was provided at Monaco by Lorenzo Bandini when he put his car on to the front row of the starting grid with a practice time beaten only by Jack Brabham; for this race he was to be joined by Chris Amon, whose planned debut as a member of the Ferrari team in the Race of Champions had been prevented when he was injured in a road accident on his way to Brands Hatch, two days before the race, when an absent-minded woman driver contrived to write off Chris's treasured Sunbeam Tiger.

As usual, only 16 cars could be accepted for the Monaco Grand Prix, and with 11 places already reserved for works cars – two each for Brabham, BRM, Cooper, Ferrari and Lotus and one for

Honda – eight drivers had to fight for only five remaining places on the grid. Gurney was easily the fastest of these with his Eagle-Weslake, although Ginther was unable to qualify the second car (he announced his retirement from grand prix driving shortly afterwards), and the others to earn a place were Bruce McLaren, who had installed a 2 litre BRM V8 engine in a modified McLaren Formula 2 chassis as a temporary expedient pending the delivery of BRM's first 3 litre V12 'production' racing engine, Jo Siffert in the Walker Cooper-Maserati, Piers Courage in a 2·1 litre BRM entered by Tim Parnell, and Johnny Servoz-Gavin, the young blond-haired Frenchman, who was driving a Cosworth FVA-engined Formula 2 Matra with additional ballast to bring it up to the minimum Formula 1 weight limit.

The Automobile Club de Monaco's invitation to Matra to send along a pair of Formula 2 cars for the Grand Prix (Jean-Pierre Beltoise was unable to qualify after his practice troubles) was a significant move prompted by the announcement a short time earlier that the French Government had offered an indefinite loan of approximately £500,000 for the design, development and manufacture of an all-French grand prix engine as a stepping stone in the climb to put France back on top in Formula 1 racing. Matra had announced their acceptance of the challenge, and indeed even then were believed to have a V12 engine on the drawing board, and no-one who had watched the very professional way in which the missile-manufacturing company had launched themselves into motor racing as car manufacturers and witnessed the performance of their ultra-competitive Formula 3 and Formula 2 single-seaters were prepared to say that they would not be able to do the same thing in Formula 1. The anticipated addition of Matra to the ranks of grand prix teams, therefore, was considered by several prominent team managers to represent perhaps the greatest threat so far to their aspirations. That the Monaco organizers had seen fit to invite Matra to take part in a two-car 'dress rehearsal' did little to alleviate their fears!

The 1·95 miles street circuit, of course, was the one venue in the world championship calendar at which an undersized car's power disadvantage would be minimized. This is why BRM once again had elected to run a 2·1 litre V8 BRM for Jackie Stewart (although Mike Spence was told to persevere with his H16), and why Colin

Chapman had provided Lotus 33s for both his drivers in preference to the H16-engined Lotus 43s, Clark having the Climax and Hill the BRM-engined car.

The Monaco Grand Prix was the one Grand Prix which Jim Clark so far had failed to win, and despite only managing a place on the third row of the starting grid due to practice problems he felt that this time he stood a very good chance. Brabham, he figured, would be the man he would have to beat, but when he saw Bandini shoot into the lead at the start and smoke begin to pour from the rear of Jack's car he knew his main adversary would soon be out of the race. Sure enough, Brabham spun on his own oil on the first lap, but Clark was also in trouble a few seconds later when he took to the escape road at the chicane after seeing the car in front of him wag its tail on another patch of oil. He was now at the back of the field, and had to start one of his mammoth comeback fights. He drove brilliantly, picking off car after car and setting a new lap record despite the extremely oily track surface, until he was challenging McLaren for third place with more than half the race still to run. But suddenly a rear damper mounting gave way as he went through the Tabac corner behind the pits. The Lotus slewed sideways, hit the barrier, and came to rest with broken suspension. Once again Clark had failed to finish a Monaco Grand Prix, far less win it.

Meanwhile Bandini, who had led the race from the first corner, had dropped behind Hulme and Stewart on the second lap, while Stewart squeezed past Hulme to lead the race during lap six. But Jackie's hopes of a repeat of his 1966 victory were dashed on lap 15 when he retired with a broken gearbox, and from then on it was Hulme all the way. This was his first victory in a world championship event, but it was a race which was to offer him, or indeed any of the other drivers at Monaco, little joy.

Lorenzo Bandini, who had invariably driven well on the testing Monaco circuit, had maintained a spirited chase in his Ferrari, and between lap 40 and lap 60 had been consistently faster than Hulme so that he had been able to reduce the gap between the two cars from about 20 to less than eight seconds. Then Hulme had responded to the pressure, and the time interval had opened up again to around ten seconds. This was motor racing at its finest, with two drivers in two very different types of car, both right on the limit and clearly very evenly matched. But whereas Hulme was able to conserve his strength and concentration, Bandini was becoming slightly ragged in his valiant attempt to give Ferrari a victory. Then, on lap 82 tragedy struck as the Italian slightly misjudged his line entering the tricky chicane. The Ferrari bounced off a barrier, slewed across the track, mounted the straw bales on the opposite side, turned over and ignited as it landed back on the track upside-down. It was several minutes before Bandini could be extricated, and he died three days later from his terrible burns. It was a sad ending to what hitherto had been a most exciting grand prix, and a painful reminder that in motor racing the margin between triumph and tragedy can be desperately narrow.

Nearly a month was to pass before the grand prix circus was to meet again at the seaside resort of Zandvoort in the North of Holland. Every world championship race generates its own characteristic atmosphere, the major ingredients of which seem to be environment, the level of friendliness of the race organizers, the enthusiasm and size of the crowd, the climate and its influence on practice and the race, and inevitably the quality of the social activities which occur throughout the period when the GP circus is in town.

Zandvoort is not necessarily so well endowed in all these aspects – the wind can blow off the North Sea with a biting ferocity, there is a severe shortage of restaurants, and the overall standard of the cuisine is notable mainly for its mediocrity – yet the atmosphere which the town generates during the week of the Dutch Grand Prix is such as to make this one of the more popular staging posts in the long journey from South Africa to Mexico.

The majority of the teams stay either at the Hotel Bouwes, overlooking the sea in the centre of the town, or at the newer Bouwes Palace, a tall structure which was erected some years ago about half a mile closer to the circuit, and which Innes Ireland was convinced would collapse on top of him on the one occasion he stayed there. Lest anyone should take the former Lotus driver's prediction seriously it should be recorded that since the day in question the hotel register has carried the names of most of the hierarchy of motor racing, not one of which has been lost by falling masonry, though several have had cause to avail themselves of the revitalizing properties provided in the hotel's sauna baths following their over-enthusiastic support of the downstairs bar!

Apart from the two hotels in question the third major oasis of activity away from the circuit itself is the garage on the main road running between them, the generous proportions of which make it an ideal maintenance headquarters for several of the teams. On the eve of the Grand Prix as many as a dozen Formula 1 cars can be seen in various stages of undress as mechanics work under floodlights in adjacent bays stripping and preparing them for the following day's race. A crowd of enthusiasts can usually be found clustering around the filling station waiting for a car to be wheeled out into the forecourt, or peering through the windows searching for another name to add to their already well-filled autograph books. But the crowd was larger, and perhaps more knowledgeable than usual in 1967, for this was the first public appearance of the Lotus 49 and the Ford DFV engine.

Even with the extra month's breathing space after the Monaco race there had been a considerable rush to get the new cars prepared in time for Zandvoort, principally because of Jim Clark's and Graham Hill's absence abroad. Between his various testing and qualification sessions at Indianapolis Hill had been rushing back to England to spend some time bedding-in his Lotus 49, but Clark was destined not to see his brand-new car until he arrived at Zandvoort on the morning of the first of the two days allotted for official practice. Mike Costin had been the very first person to drive the new car and to sample its 400-horsepower engine, and Colin Chapman's racing manager and former chief mechanic Dick Scammell had also slipped on a crash helmet to help speed up some of the routine test driving at Hethel. But the serious tailoring of a car to its driver can only be carried out by the driver himself, and although Graham Hill was able to complete much of this work before his car left for Holland, when Clark first sat in his car at Zandvoort it was still 'factory fresh'. Consequently, he had to spend much of the first day's practice getting the car to fit him and then acclimatizing himself to its controls, its chassis responses and to the characteristics of its engine which were, to say the least, unusual in that there was a sudden upsurge of power around 6,500 r.p.m. This meant that you had to be most careful in your choice of engine revs through all the corners.

As so often happens with a new car the Lotuses proved to be wrongly geared for the circuit, and both cars were fitted with replacement gearboxes with more suitable ratios during an interval during practice sessions on the first day. At this point the fastest laps had been turned in by Hulme in his Repco Brabham and Rindt in his Cooper-Maserati, but Hill was easily the fastest of all during the final session that day, a full second quicker than Surtees in the Honda, while Clark was within a fifth of a second of Surtees' time. So far the Ford V8 engines seemed to be standing up well, apart from a temporary electrical problem which had delayed Hill during the afternoon, but Clark was not too happy with the handling of his car.

That night his suspension was checked and re-checked, but no fault was discovered apart from a small amount of slack in the tapered wheel bearings which was eliminated. But the real problem was revealed soon after the start of the final qualifying session on the Saturday when Clark returned to his pit with a ball joint broken in his right rear suspension, the failure having also fractured the hub carrier. After a careful examination confirmed that it had been the faulty ball joint and not a sub-standard hub casting which had caused the trouble, a replacement suspension was made up in the paddock, but by the time it had been fitted to Clark's Lotus practice was nearly over and he had no chance to improve on his Friday lap time which was now only fast enough for a place on the third row of the starting grid. Hill, however, was making real progress with his car, his times coming down progressively until only Gurney in the Eagle-Weslake remained a strong challenger for pole position. A few minutes from the end of practice it looked as though the Californian had made it with a time of 1 minute 25·1 seconds, a full three seconds faster than Brabham had managed in practice the previous year. But Hill went out again, equalled Gurney's time on two successive laps, and then finally put in a blistering last lap which several people timing in the pits thought was 1 minute 24·9 seconds, but which the official timekeepers announced as 1 minute 24·6 seconds. Either way, the Lotus was undisputably the fastest in practice, and there was restrained jubilation within the team, and more abandoned excitement by the large contingent of Lotus and Ford supporters who had gathered at the circuit but who knew little of the tensions which build up in the paddock and in the garages between final practice and the race itself.

There must have been many crossed fingers in Zandvoort that evening for a number of reputations rested on what would happen the following afternoon. It may have been unrealistic and unfair to expect the new car and engine to win their first race, but nevertheless this is what the public wanted to see and indeed was now expecting to see. To have secured pole position first time out was something for the record book and a very considerable achievement, but in some respects it was also an embarrassment in that inevitably it focused the pre-race spotlight firmly on the Lotus 49 and its engine, and this meant that anything less than a victory in the Grand Prix would be something of an anti-climax.

In times like this it helped to have something to do, and there was certainly a lot to do back at the garage by the Team Lotus mechanics and the various technical and competitions staffs of the supporting component suppliers. Keith Duckworth scrutinized his engines with characteristically meticulous care while Colin Chapman stood by as the two cars' suspensions were examined minutely to ensure that there was no repeat of the ball-joint trouble which had robbed Clark of so much practice time.

While this work went on well into the evening, two people with particular cause to hope for a miracle the following afternoon, Walter Hayes and Harley Copp, were driving towards Amsterdam for a dinner date at the home of Ted Edwards, the Managing Director of Ford in Holland. Hayes remembers that evening as one of unbelievable tranquillity, the excellent meal being followed by Copp, who is an accomplished classical pianist, entertaining his colleagues in the drawing room with a selection of airs which seemingly had a magnificently therapeutic effect on their eve-of-race tensions. 'It was all desperately relaxed, though I have to admit that we were not so relaxed the following morning.'

They lunched early in Zandvoort with Raymond Mays of BRM and Jim Elwes, a Ford main dealer from Sussex, then made their way to the circuit through the tunnel beneath the main straight, and on into the paddock. There was still some time before the cars were due to be marshalled out on to the track, but already the pre-race tension and exaggerated activity was in evidence as people rushed around searching for that spanner that was sticking out of their overall pocket, polishing that spare pair of goggles for the third time, and stopping and re-starting that set of stop-watches to make sure that they were still working as well as they had been five minutes previously when they were last checked. This is all part of the familiar paddock scene before every important race.

Colin Chapman busied himself checking and rechecking everything in sight and shouting orders to anyone who would listen to him (and several people who wouldn't!); Keith Duckworth tried hard to convince himself that there was no point in worrying any more, but failed; Jim Clark was easing himself into a state of introversion which so often had preceded one of his great drives; Graham Hill had long since passed through the time barrier beyond which it was dangerous to attempt to engage him in polite conversation; Harley Copp sought distraction with the aid of his movie camera; and Walter Hayes once again indulged in his mime of the toilet soap commercial as he walked around the paddock telling everyone with a press pass that the engine had already shown its worth in practice, and that if necessary he was willing to wait several grands prix for it to win its first race. He was talking good sense, of course, but nobody would believe him!

Eventually it was time for the cars to leave the paddock and for the drivers to climb aboard for a warming-up lap before assembling on the starting grid. Hill had the favoured right side of the front row, which would give him the inside line entering the first corner provided he had not been passed by Gurney or Brabham, who were starting to the left of him. But Clark was less well off, having the left side of the third row, with Hulme and Surtees to his right, and Rodriguez and Rindt in the row in front.

What happens during the first few seconds of a race can easily determine the final result because this is the one occasion on which you can pass a bunch of cars simultaneously and so cancel out a poor grid position. Conversely it is also the time when an excess of wheelspin or a missed gear can cost you several places. The determination to make a good start, therefore, is uppermost in every driver's mind at this time, though coupled with awareness that the slightest error of judgment in such heavy traffic can be disastrous.

Both Hill and Clark were usually better than most at getting their cars off the line, and wisely both had practised a standing-start so as to familiarize themselves with the power curve of the Ford engine. After all 17 cars had been eased forwards from the assembly grid on to the final starting grid, the starter raised the Dutch national flag, swept it down and took a sharp pace backwards as the pit area was suddenly covered by rubber smoke. Hill and Brabham made the best starts from the front row as Gurney lagged slightly and was passed by Rindt, while Clark made a great start from the third row and was only half a length behind Gurney as they swept into the Tarzan Corner. Hill had won the acceleration race and was first through the 180-degrees turn, and as the cars swept past the back of the pits Graham was being chased by Brabham, Rindt, Amon (who had made a tremendous start from the fourth row), Gurney and Clark.

By the end of the first lap Gurney had repassed Amon's Ferrari into fourth place and Hill had already pulled out a lead of over a second from Brabham. The race was to be over 90 laps, and at the end of the first ten Hill was out on his own, dictating the pace, while Brabham, Rindt, Clark, Hulme and Amon were fighting closely for the next places some way behind. Gurney's race had already ended with a broken fuel metering unit so that one of the biggest threats to the Lotus-Fords had been removed.

However, the next time round Brabham was leading the race and there was no sign of Hill, who arrived much later pushing his car towards the pits. The engine had just stopped, and so the plugs were changed in the hope of reviving it, but it was soon obvious that the trouble was more serious than that, and the fault was eventually traced to a timing gear failure. Now it was all up to Clark whose car was still not fully sorted-out, but who at least had an engine that had covered many fewer miles than Hill's, which had been used for much of the development testing back in England.

Seeing his team-mate at the pits Jimmy started to pile on the pressure and overtook Rindt into second place on lap 16, and on the following lap he went past in the lead with Brabham trailing him closely. There were still 74 laps to go, which meant the best part of two hours' racing, and the tension in the Lotus pit was soon becoming unbearable.

At 20 laps Clark was leading Brabham by just over a second-and-a-half, while Hulme had moved up behind his team-leader and Amon was right behind him in fourth place. The race was now between these four as the remaining cars had been dropping back gradually and had now lost contact. At 40 laps Clark had extended his lead to 12 seconds, and at the 50-lap mark the gap was up to nearly a quarter of a minute. Every time the Lotus passed the pits ears were strained for any tell-tale signs suggesting trouble, but although the engine occasionally popped and banged on the overrun – a characteristic which was to become a familiar sound during the 1967 season – Clark's lap times remained constant and his lead continued to increase. By lap 60 he was about 25 seconds ahead of Brabham, and ten laps later the gap was up to 37 seconds.

With only 20 laps to go Clark was driving as faultlessly as ever, and both car and engine seemed to be in perfect shape. But Chapman was taking no chances and he ordered the 'easy' signal to be shown to Jimmy, who immediately responded by dropping his revs and lengthening his lap times. As a result the engine's note changed, and the huge crowd in the grandstands and on the sand dunes lining the circuit, who had been willing Clark on to his fourth Dutch Grand Prix victory, sensed that he was suddenly in trouble. But they need not have worried because he swept on to complete his 90 laps, and lead Brabham across the line by a comfortable margin of 23·6 seconds.

The Lotus 49 and the Ford DFV engine had won their first race at the first attempt, and the release of tension in the pits was so great that pandemonium reigned for minutes afterwards. Clark was mobbed by every Lotus and Ford man in sight when he arrived back at the pits after his cooling-down lap, was released temporarily so that he could climb the victor's rostrum and collect his garland, and then was mobbed again in the paddock where the celebrations went on long after the race.

Jim Elwes rushed up to Walter Hayes and said, 'This is the greatest day of my life', and altogether it became a very emotional scene. The celebrations were to be continued throughout the evening in Zandvoort, first at the official prizegiving party and later in all the bars and restaurants which do such a roaring trade during the night after the Grand Prix, but long before then two of the architects of the victory had already left for home.

Walter Hayes and Harley Copp had decided that it should be Jimmy's and Keith's celebration, and that it would have been very undignified for them to stand around and crow. So, while the champagne and the beer was still flowing freely in the paddock, they had climbed into their car and slipped away towards Schipol airport near Amsterdam, where the Ford executive aircraft was waiting for them. Along the way they stopped the car outside a shop and bought two bottles of beer which they drank in a private toast to each other as they continued their way to the airport. While the Zandvoort revelry was still at its height they were back in England, feeling completely shell-shocked. They had been hoping for a miracle, but somehow they had been not quite prepared for it.

7

A message from the sponsor

THE DUTCH GRAND PRIX had provided a story book beginning to the racing history of the Ford DFV engine, but the Zandvoort victory, though a justifiable cause for jubilation, was in no sense any reason for complacency. The engine had achieved its target of 400 horsepower with something to spare, and this had proved more than adequate to dispose of all current opposition. But the failure of Hill's engine fairly early in the race proved that there was a certain amount of development work still to be done, while the ultimate ease of Clark's victory was expected to act as a considerable spur to the opposition, who could be relied upon to mount a counter-attack, and mount it quickly.

On the evidence provided during practice at Zandvoort, the Eagle-Weslake still seemed to offer the greatest potential threat, and it was significant that Gurney had been able to set the second-fastest practice time on a circuit which might have been thought to favour the Lotus more than the Eagle. On the faster circuits, such as Spa and Monza, the V12 engine might take an awful lot of catching provided it remained healthy, and for this reason there was more than usual interest in the outcome of the Belgian Grand Prix, which was to take place two weeks after Zandvoort.

Dan Gurney arrived at Spa-Francochamps with his ever-present grin wider than ever, for during the intervening weekend he and A. J. Foyt had taken their 7 litre Ford GT to a record-shattering victory in the 24 Hours of Le Mans, a success which so delighted them that they had proceeded to spray each other and everyone else in sight with their victors' champagne during the post-race celebrations!

But Gurney's cause of contentment was not confined to his Le Mans success for things were beginning to look up for him in Formula 1 as well. The Weslake engine was said to have recorded an impressive 416 b.h.p. at 10,000 r.p.m. on the test bed, and its durability was improving all the time. The lanky Californian naturally was hungry for his first grand prix success with his own car, and he was soon setting some impressive practice times on the very fast but very testing Spa-Francochamps circuit. Throughout both days of practice he was the only driver to offer any serious threat to Jim Clark's fastest lap with the Lotus-Ford, and he lined up in the centre of the front row for the Grand Prix, with Clark on his left in pole position and Hill on his right.

Jimmy was now feeling a lot happier with the Lotus 49, which had been strengthened around the rear of the monocoque after the Zandvoort race, and been fitted with a brake-balance control in the cockpit. During the final training session he had put in a sensational lap in 3 minutes 28·1 seconds, which was 3·1 seconds faster than Gurney's best, and 4·8 seconds quicker than Hill's time, which had been recorded on the first day. It looked as though the race would be between these three as Rindt and Amon, in their Cooper-Maserati and Ferrari respectively, had only managed 3 minutes 34·3 seconds for their second-row positions, and Stewart in the H16 BRM and Brabham in his latest lightweight BT24 – based on his Formula 2 car – were the only other drivers to bring their times below 3 minutes 36 seconds. But practice times are not always a reliable guide to what will happen in a race.

Lotus were in trouble even before the flag fell when Hill's car refused to start and he was left on the assembly grid as the remaining 17 starters eased forwards on to the final starting grid. After the flag had been dropped Hill freewheeled downhill to his pit, had a replacement battery fitted, and eventually got away when the remainder were already over half-way through their first lap, but he was to retire with a broken clutch after only three lonely laps.

Clark, however, was soon proving that his practice lap was no fluke by building up a large lead while Gurney, who had made a hesitant start, had dropped behind Stewart, Amon and Rindt on the first lap. Then Amon spotted the wreckage of Parkes' Ferrari which had crashed badly on the opening lap, and fearing a repeat of the Bandini tragedy (in fact Parkes escaped, but his injuries were to keep him out of racing for two years) the New Zealander lost his concentration and began to drop back.

Gurney, meanwhile, had recovered to third place behind Stewart, but there seemed nothing that either of them could do to stop a runaway victory by Clark. But then, with 12 of the race's 28 laps completed, the whole complexion of the Grand Prix altered as Clark rushed into his pit with a misfiring engine, to be followed a few seconds later by Gurney, who had been on the point of snatching second place from Stewart, who was now in the lead with the big BRM.

Clark's trouble was traced to an over-tightened plug, which had broken its electrode, and after it had been removed and replaced Jimmy went on his way again, but by this time he was way down in ninth place and out of contention. Gurney's stop, meanwhile, had been very brief. He had called in to report that his fuel pressure was fluctuating between 60 and 120 lbs per sq in instead of its usual steady 145 lbs, but team manager Bill Dunne urged him to carry on racing as there was nothing that could be done quickly.

The stop put Gurney nearly a quarter of a minute behind Stewart, but immediately he started to eat into the lead, at first slowly, and then much more rapidly as Stewart ran into gear-selector trouble and had to drive one-handed to keep the car in gear; once or twice when he missed gears his revs shot up to over 12,000 r.p.m., but somehow the engine managed to hang together.

Gurney was now closing in at the rate of four seconds a lap, and the Eagle went through into the lead as the two cars swept downhill from La Source hairpin past the pits at the end of their 21st lap. But the jubilation in the Eagle's nest was tempered by anxiety by the growing cloud of smoke which was appearing out of the back of their car. As it transpired there was no cause for alarm because it was caused merely by oil dripping from a breather pipe on to a hot exhaust, and Gurney was able to motor on to a comfortable victory, over a minute ahead of Stewart's BRM, which in turn was over a half a minute in front of Amon's Ferrari. Poor Clark had to make a second stop with similar plug trouble, and later on was left with only third and fifth gears, so that in the circumstances he did manfully to finish sixth behind Rindt and Spence (BRM), after completing 27 laps. So, just one week after winning the fastest-ever Le Mans race, Dan Gurney had won the fastest-ever grand prix, his average being almost 146 m.p.h. and his fastest lap of 3 minutes 31·9 seconds, the new official record, setting an average of 148·85 m.p.h.

There could scarcely have been a more vivid contrast between the ultra-fast sweeps and dips and climbs which characterize the Spa-Francorchamps circuit and the tight corners and short straights which combine to make up the short Bugatti circuit at Le Mans, which was to be the unlikely scene of the French Grand Prix in 1967. It was a travesty that the oldest GP in the international calendar should have been staged on a circuit more suited to Formula 3 cars, or to the racing drivers' school whose permanent home it was. Nevertheless, the Automobile Club de France decreed that it should be so, and the Automobile Club de l'Ouest, who operate the facilities at Le Mans, looked forward optimistically to their second vast crowd of enthusiasts to the circuit in the space of three weeks. But in this they were to be bitterly disappointed, for an estimated 20,000 people were to turn up for the Grand Prix, or about 10 per cent of the crowd which habitually fill the stands and terraces for the 24-hours race. For the remainder, the sight of Formula 1 cars racing on a circuit consisting of the pit straight and first corner of the endurance race course, plus a series of twists and turns behind the car park area beyond the paddock, was simply no attraction. The events of the day were to prove that their money was well saved.

It was a blisteringly hot day as a small group of 15 cars was assembled on the starting grid, the regular field being depleted by the reduction of the Ferrari team to a single car for Amon and the absence of Surtees' Honda, which had blown up early in the race at Spa. McLaren was taking his first drive as the second member of the Eagle team, his own 'oversize Formula 2 car' having been damaged at Zandvoort and subsequently discarded as uncompetitive (he was still waiting vainly for his promised BRM V12 engine). The Le Mans race, therefore, seemed to rest between the two Lotus-Fords of Clark and Hill, the two Repco Brabhams of Brabham and Hulme and the two Eagle-Weslakes of Gurney and

McLaren, with possible intervention by Amon's Ferrari and Rindt's Cooper-Maserati.

Practice, which had put Hill, Brabham and Gurney on to the front row of the grid in that order, had demonstrated the futility of running the GP on the Bugatti circuit. The course was so tight around the back that there was virtually no chance of any passing there, while the hairpin which led back on to the pits straight slowed the cars so much that as they passed the pits accelerating uphill towards the Dunlop Bridge they looked like racing cars in slow motion.

The race was to be over 80 laps, and at the end of the first Hill, who had made a very good start, led from Gurney, Brabham, Clark, Amon and McLaren. Clark had been in persistent fuel pump trouble during practice, which explained his second-row grid position, but his replacement engine seemed to be functioning much better, and he moved up to take the lead on lap 5, while Brabham went ahead of Hill, though only temporarily. By lap 7 the two Lotus-Fords were running first and second, and gradually they opened up a gap on Brabham, Gurney and Hulme while McLaren ran into an ignition problem and dropped back with the second Eagle-Weslake.

For several laps Clark and Hill ran nose-to-tail, then they exchanged places, Hill going ahead on lap 11, but three laps later Clark reappeared on his own. Hill's car had come to a stop around the back of the circuit, with his crownwheel and pinion stripped of its teeth. Clark then settled down to a long and lonely job, keeping a comfortable distance ahead of Brabham, Gurney, Hulme and Amon, but on lap 23 he, too, ran into transmission trouble and headed for the Lotus pit, his race over. The constant gear-changing and sudden changes from acceleration to deceleration had proved too much for the Lotuses' transaxles, the cases of which had flexed, allowing the gears to run out of mesh and eventually to strip. So if nothing else the Bugatti circuit had at least served to pinpoint a transmission design weakness which ZF were able to rectify by cross-bolting the cases in time for the next race.

With both the Lotus 49s out of the race the way was clear for a Brabham victory, and when Gurney dropped out on lap 41 with a fractured union in the fuel injection system, Hulme moved up to make it a clear-cut Repco Brabham one-two. Stewart, who gambled on using one of the old 2 litre V8 BRMs in place of his regular 3 litre H16, proved the wisdom of his choice by coming home in third place, one lap in arrears. The only consolation for Team Lotus was another lap record by Graham Hill, while for Jim Clark the struggle to take the world championship in 1967 was beginning to look a little grim, for Hulme's second place had hoisted his points total to 22, while Brabham had moved up to take second place with 16; so far, Clark had been able to muster only 10. But the season was certainly offering plenty of variety, for the five championship races so far had produced five different winners in four different makes of car.

The openness of the championship battle, and the variety in the race results heightened pre-race interest in the British Grand Prix, which for 1967 was to take place at Silverstone. There was a strong contingent of spectators at the circuit, therefore, to watch practice on the eve of the race, and those who gathered in the grandstand at Woodcote Corner, just before the pits, were to witness an unexpected piece of drama.

Everything had been going very well for Team Lotus, with Jim Clark making fastest time, Graham Hill second fastest, and Brabham, Hulme and Gurney their only close challengers. Then, with less than an hour of practice to go, Hill decided to return to his pit after setting a couple of fast laps. He entered the pit approach road about 70 m.p.h., applied his brakes, and suddenly the Lotus turned sharp right and demolished itself against the retaining wall. Bits of car and wall flew everywhere, and an astonished Hill hopped out to retrieve a detached wheel which was rolling down the track in the path of oncoming cars.

Returning to his car he found a sorry-looking sight. The front right wheel, suspension and radiator had been ripped away from the badly torn chassis, and it was very apparent that the car was out of the race for it would be impossible to repair it quickly, if at all. The cause of the accident was to be found at the back of the car. A faulty weld securing the lower left radius arm to the chassis had broken, and the detached arm had allowed the left rear wheel to suddenly turn left, which had the effect of steering the car sharp right into the wall.

The Lotus mechanics rushed from their pit to retrieve the car as practice was abandoned temporarily, and Colin Chapman examined the wreckage thoroughly. As only two Lotus 49s had been built so far it seemed certain that there would be only one of them – Clark's – on the starting grid tomorrow. But Colin had other ideas. Back at the Lotus factory there was another bare chassis structure. It had nothing fitted to it in the way of suspension or any other running equipment, but at least it had been wired and plumbed. So Chapman rounded up every available Lotus mechanic and ordered an all-night session for a new car to be built in time for the race. What could be salvaged from the crashed car would be stripped off and fitted to the new chassis, and everything else would be taken off the shelf or if there were no parts available they would be made up from scratch. To add to the problem, the third chassis differed in many small but important respects from the previous two, and this meant that many of the parts from the earlier car would not fit the new monocoque, at least not without adaptation.

Work on building up the new car began within a couple of hours of the accident as Chapman had flown several mechanics straight from Silverstone to Hethel, leaving one behind to drive the wreck and Clark's car back in the transporter. By the time a few phone calls had been made to employees who thought they were enjoying a weekend off, a total of 16 people had been alerted for the job. Normally it takes three weeks to build up a new car, but Lotus almost managed it overnight.

In fact they were still working on the car as it was driven back towards the circuit, and the red-eyed mechanics were given a great cheer when they unloaded the car in the paddock. But there was still work to be done, and it continued right up to the time when the cars lined up on the starting grid. Also, the work was not all confined to Hill's car, as Clark's also had to be thoroughly examined and prepared for the race, and as a precaution the rear suspension brackets on both cars were reinforced to prevent the possibility of a similar breakage.

It was a miracle that two Lotuses lined up on the front row of the grid alongside the two Brabhams, and no-one was unsporting enough to protest against Hill's place on the front row when technically he should have been at the back of the grid because he was driving a car other than the one which he had used for practice (even though it had many of the same components).

There was almost a fairytale ending to the story, because Clark rocketed off into the lead, and after Hill had disposed of Brabham he settled down into second place, closed on Clark, and then took over the lead on lap 26. The race was over 80 laps, and Hill was still in first place on lap 54, with Clark a short distance behind him.

But going into Beckett's Corner on lap 55 Hill's Lotus suddenly weaved all over the road in a most alarming fashion, and after regaining control and negotiating the corner Graham looked around to find his left rear wheel inclined inwards. It was a long way back to the pits, and he had to make the journey very slowly so that by the time he drove up the pit ramp he had lost over a lap.

Chapman took a quick look at the drunken wheel, found that an Allen screw securing the inner end of the top suspension link had fallen out (it had probably been insufficiently tightened in the mad rush); a replacement was found and fitted so quickly that the car was at rest for barely a minute. But Hill was now two laps in arrears, and although he pressed on gallantly, hoping to make up some of his lost time, his engine blew up eight laps later, and he walked back to his pit to receive an ovation from the crowded grandstands. But the applause was as much for the Lotus mechanics as for Graham on this occasion, and when he was interviewed a few minutes later he paid them generous tribute for their gallant work. But all was not lost for Lotus and Ford that day because Clark went on, completely unchallenged, to win his fifth British Grand Prix from Hulme, Amon and Brabham.

Forward planning is essential to the good conduct of most businesses, and it is never more vital to success than in motor racing. It was now time, therefore, to decide on the future of the DFV engine. Up to now Lotus had had sole use of the engine, and this situation would remain until the end of the year. But what of 1968? There were strong arguments against continuing the exclusive arrangement with Lotus for another year from both the Ford and the Cosworth points of view.

The Ford Motor Company's involvement in grand prix racing, although initially involving a tie-up with one team, was given the corporate go-ahead because it was felt that this was a method by which Ford could help grand prix racing as a whole, and through fostering British success in it, could ensure the continuity of prestige which Britain's motor industry had been enjoying through the country's racing achievements. Whilst there seemed little doubt that the Lotus 49, with its Ford engine, would start a favourite to win most grands prix, by maintaining an exclusive agreement with Team Lotus for a further year Ford might well be considered to be doing grand prix racing a disservice by minimizing the chances of other British-based teams which had expressed a keenness to use the DFV engine. Indeed, there was more than a possibility that one or two teams would withdraw from Formula 1 racing if the Ford engine was not made available to them.

The Cosworth desire to serve additional teams was partly because the company now had the facilities to manufacture and maintain more engines than would be needed by Team Lotus, and partly because they did not wish to become styled simply as 'Purveyors of engines by appointment to Lotus'. They had had close and very amiable links with Lotus in a number of their engine building operations, and whilst this was an important and welcome part of their business they did not want to give anyone the impression that they were tied to one car manufacturer, whether it be Lotus or any other company. Consequently, they favoured strongly a freer market, subject to their right at least to have a say in the choice of customer for their additional engines, which obviously would have to be limited in number.

The news was broken to Colin Chapman at the Lochmühle, a delightful hotel some 25 miles from the Nurburgring, when Colin sat down for dinner with Walter Hayes, Keith Duckworth and Walter's son Jeremy, on the eve of the German Grand Prix. Naturally, Chapman was disappointed at the news, but he understood the logic of the arguments put forward, and the informal meeting ended on a very friendly note with the Lotus chief assured of fairly substantial support from Ford during the 1968 season (his engines would continue to be supplied free of charge) and an undertaking that no more than 15 additional engines would be built for other customers, who would be required to pay £7,500 each for them. At about this time Keith Duckworth let it be known that one day he might talk himself into building his own Cosworth grand prix car around the DFV engine, and Walter Hayes on hearing this indicated that should he decide to pursue the idea the Ford Motor Company might well be prepared to offer some backing.

One way or another, the German Grand Prix meeting was to provide Colin Chapman and his drivers with one of their most depressing weeks of the year, for scarcely anything went right for them at the Nurburgring. Hill's car seized up during his first lap of practice because there was no oil in the gearbox, and after making the long journey back to the pits he climbed into the spare car, and after a few laps had a most horrifying accident in it through going into a corner too fast. He had been experiencing some brake trouble and was unable to slow the car down sufficiently with the result that he ran off course, up a bank, and performed a 'wall of death' act before landing back on the track with his very bent car travelling smartly backwards.

Graham had to borrow Jimmy's car for one practice lap to bring his quota of training laps up to the minimum permissible five, and he was well back on the grid for the start, which probably helped to promote his incident before the very first corner of the race. The Grand Prix that year was open to Formula 2 as well as to Formula 1 cars, the smaller cars being marshalled into separate rows immediately behind the Formula 1 grid. On the rush away from the start some of the Formula 2 cars became mixed up with the Formula 1 tail-enders, and Hill suddenly found himself being nudged in the rear as he braked for the South Curve. With wheels locked up he disappeared off course in a cloud of rubber and brake smoke, and after the dust had settled he was very relieved to find himself stationary, still on level ground and with the car apparently undamaged. He set off again well behind the field and was still a long way behind the leaders a few laps later when he suddenly found the car weaving all over the road as he went through a fast corner. Then he noticed that a front wheel was wobbling, and after gathering everything under control he drove slowly back to his pit where the wheel's central securing nut was tightened. He went off again, by now feeling considerably shaken, but stopped for good two laps later with a broken rear suspension.

Jim Clark's race had already ended at the pits. He had started in pole position after an electrifying practice lap in 8 minutes 4·1 seconds and had stormed straight into the lead from Denny Hulme and Dan Gurney. But on the fourth lap Jimmy collected a slow puncture, and as his car dropped closer to the track part of the front suspension started to break up as it scraped against the undulating surface, so that he was only just able to limp the Lotus back to the pits.

Gurney had moved ahead of Hulme as Clark ran into trouble, and it looked as though the Eagle-Weslake was about to repeat its Belgian Grand Prix triumph, but with less than three of the 15 laps to go one of the Eagle's universal joints collapsed and Gurney's race was run. As Hulme went on to a clear-cut victory ahead of Brabham it began to look as though the 1967 world championship would be fought out exclusively between these two team-mates, with the odds becoming steadily greater in favour of the New Zealander, who by now had collected 37 points to his boss's 25.

The Brabham domination continued three weeks later when, for the first time, a world championship Formula 1 race was held in Canada. 1967 was the country's centennial year, and with Expo 67 in full swing and attracting thousands of visitors to Canada it was the perfect opportunity to stage a championship-status race, although by sanctioning a date in August the FIA were committing the teams to two Transatlantic return trips, and all the expense that this involved, in the space of just over a month. All subsequent Canadian Grands Prix have been slotted into the busy racing calendar between the Italian and United States Grands Prix, so that teams can complete their European programme and then take in the final three races in Canada, the United States and Mexico in one long trip.

Although Canada was in a gay mood in 1967, the Mosport circuit north of Toronto was anything but cheerful on race day. The weather had been kind during practice, and the dry track had enabled Clark and Hill to show the superiority of their Lotus-Fords in convincing fashion. Hulme had proved to be the closest challenger, and he filled the third place on the front row of the grid, but one of the surprises in practice had been the performance of McLaren, who was back at the wheel of one of his own cars.

Bruce's patient wait for his BRM V12 engine had been rewarded at last, and the M5A chassis which he had built to carry it looked to be a very workmanlike job, and clearly it was handling very well indeed, despite the limited amount of testing it had been given before being shipped across the Atlantic. The engine, which had been designed originally as a sports car power unit, was reputed to be giving a relatively modest 370 b.h.p., but it seemed to have good torque, and McLaren's best practice lap was only just over a second slower than Clark's pole position time.

Clark had had a spot of bother during practice, slightly damaging his car when he hit a bank, but the damage had been easily repairable, and Team Lotus were able to hand over their spare car to a local driver, Eppie Weitzes, to use in the race. Since their Nurburgring problems, the Lotus 49s had been given new and stronger front suspension arms, and some modifications had been made to the fuel-feed systems in order to improve pick-up.

There was an ominously grey sky over Mosport on race morning, and sure enough rain was falling steadily as the cars lined up for the start. One or two had spun on the warming-up lap, which perhaps was a reminder that grand prix drivers had had comparatively little wet-weather experience all year (every 1967 GP so far had taken place in the dry), but it also indicated that Mosport in the wet was a very slippery place indeed.

Clark's Lotus emerged first out of the spray at the start, while Hill, Hulme, Stewart in his BRM and Gurney in his Eagle did their best to remain in contact. Hulme tried driving along the edges of the track to obtain more grip, and managed to pass Hill. Then he closed on Clark and moved into the lead as Brabham eased Hill back to fourth place. But McLaren was proving the revelation; after an early spin had dropped him to twelfth place he repassed seven cars in four laps, then worked his way through into second place. But just as he was attacking Hulme the rain eased. Clark went ahead again as the track dried, but Hulme and McLaren counter-attacked during a later rain storm until Hulme had to stop for replacement goggles and poor McLaren for a replacement battery; Bruce went on to finish a gallant seventh.

Meanwhile, Clark had fought off an attack by Brabham, who had inherited second place during Hulme's pit stop, but then Jimmy's engine fell victim to the rain and he dropped back with soaked ignition. He made a pit stop in the hope of drying things out, but within a lap of restarting the engine was misfiring as badly as ever again, and he had to give up altogether. As the Eppie Weitzes Lotus had been disqualified for receiving a push by mechanics when his engine had refused to restart after a pit stop, only Hill's car remained in the race. Fortunately, Graham's engine had been unaffected by the rain, but he had lost time with a spin during which the engine had stalled and refused to fire again. Hill had managed to struggle out and push the car along an almost level piece of track, then jump in, find his pedals, and select first gear, all before the car came to rest again – quite an achievement with a contemporary-style grand prix car which you tend to put on rather like a sleeping bag!

Hill's run-and-jump performance inevitably lost him a considerable amount of time, and he was now back in fourth place behind Brabham, Hulme and Gurney. Hulme, of course, was desperately keen to catch and pass Brabham in order to extend his lead in the world championship table, but while Jack continued to lap regularly, seemingly unaffected by the rain, Denny had to make two more stops to replace misted-up goggles, and was over a minute behind at the end of the race. Brabham's performance that day was eloquent testimony to his coolness when driving under adverse conditions. Most of the drivers in the race had become very tensed up by the tricky conditions, and their perspiration had contributed to the misting-up of their goggles; at one stage in the race the pit road had looked more like an open-air accessory shop as goggles and vizors were passed rapidly between cockpit and pit counter. All except Brabham; as ever there was not a sign of perspiration to be seen on his face at the end of his long and gruelling race, just the broad grin which goes with the chequered flag and nine world championship points!

The Italian Grand Prix is invariably one of the highlights of the Formula 1 season, if only because the long straights and fast curves of the Monza autodromo invite the type of wheel-to-wheel slipstream battles which send countless thousands of Italians into ecstacies of delight. Their enthusiasm, which builds up steadily throughout the days of practice, becomes contagious, so that by race time there is a carnival atmosphere at Monza and a tension which is matched in very few other places on the motor racing map.

The 1967 race had the added appeal in that not only was it carrying the courtesy title of *Grand Prix d'Europe*, which passes from one race to another year by year, but it was also marking the debut of two interesting new cars. Ferrari had a new lightweight chassis for Chris Amon, with a new V12 engine with four valves per cylinder to go with it. The engine was reputed to be delivering

408 b.h.p., so that on paper at least the new Ferrari, which was only a fraction over the minimum weight limit for Formula 1 cars, should have been fully competitive with the Lotus-Fords.

There was also a brand new Honda. John Surtees, who had missed the Canadian Grand Prix, had brought Eric Broadley of Lola Cars together with one of the Honda engineers to design and construct a completely new chassis in the space of six weeks. Inevitably, the monocoque structure was very much along Lola lines, and in fact it had a genuine Lola nose section, but a special frame had to be built up at the rear to take the V12 Honda engine. Compared with some of the other cars, notably the Lotus and Brabham, the hybrid Honda (which one or two people christened the 'Hondola'!) was not exactly a lightweight, but it was still some 200 pounds lighter than the monster cars with which Surtees had been persevering during the first half of the season.

There were also one or two driver changes to heighten pre-race interest, particularly the departure of the popular Italian driver Ludovico Scarfiotti from Ferrari, with whom he had not been on the best of terms of late, in order to drive the second of the Eagle-Weslakes. The race was also to see the Formula 1 debut of Jacky Ickx, who had been brought into the works Cooper-Maserati team as a replacement for Pedro Rodriguez, who had damaged a foot badly in a Formula 2 accident several weeks earlier.

There was a certain amount of chaos surrounding the electronic timing during practice, and one complete set of times had to be ignored. There was also a certain amount of controversy concerning one or two of the final official practice times, on which grid positions were based, but as positions chop and change from lap to lap at Monza as people weave in and out of each other's slipstreams, grid positions are not necessarily so vital there. The most important thing is to make a sufficiently good start to ensure that you earn a place somewhere in the first bunch of cars during the early laps, however large that bunch may be.

As it transpired, the start of the race was to be even more chaotic than the practice timing because the person who signalled the cars forward from the assembly grid on to the final grid waved his flag so enthusiastically that half the field decided there and then that the race was on, and went screaming off in a cloud of wheel-spin, to be followed almost immediately by the remainder who had been taken by surprise. The starting grid order, therefore, of Clark, Brabham, McLaren, Amon, Gurney and Hulme bore comparatively little resemblance to the first lap order of Gurney, Brabham, Hill, Clark, McLaren and Stewart.

Gurney's lead in the Eagle-Weslake was short-lived because he broke a connecting rod and drove slowly back to his pit on the fifth lap, by which time Clark had worked his way back into first place ahead of Hill, Hulme and Brabham. Gradually the leading quartet broke away from a further four drivers – Surtees, Amon, Stewart and Rindt – who engaged in a close battle for fifth place, and the remainder of the field, including Giancarlo Baghetti in the third Team Lotus 49, were slowly losing contact with the leading battles.

Then Hulme and Brabham noticed that one of Clark's tyres was deflating, and they squeezed past as quickly as they could to warn him of the trouble (on a high-speed circuit such as Monza, when you are travelling at between 150 and 180 m.p.h. for much of the time, you get little warning of trouble of this sort as the centrifugal force of the wheel tends to keep the tyre away from the wheel rim, but of course you can be in serious trouble when you slow for a corner and the car becomes unstable).

Clark was thankful for the warning, and he made his way back to his pit for a wheel change, coming out again to rejoin Hill, Brabham and Hulme just as they were completing their next lap. His stop had dropped him right down to 15th position, but as there were 54 of the 68 laps still to run Clark was determined to fight his way back through the field again. With Hill's assistance he put in a big effort and managed to shake off the two Repco Brabhams, and soon the two Lotus-Ford drivers were lapping the mid-field men, which meant that as Hill built up a commanding lead in the race Clark was pulling himself steadily up the leader board.

At the half-distance mark Hill was nearly a quarter of a minute clear of Brabham and Hulme, while Clark was back up to seventh place with Rindt, McLaren, Amon, Surtees and Brabham still ahead of him, Hulme by this time having dropped out with engine trouble. Baghetti was having a good drive with the third Lotus, and was just one place behind Clark, and so despite Jimmy's puncture it looked as though Lotus were heading for a good all-round result. Baghetti's remarks after having first tried the Lotus during practice had been an amusing commentary on a driver's

The master at work. Keith Duckworth often sat at his drawing board from breakfast time to midnight designing and detailing the Ford engine.

The end product – a superbly compact V8 power unit, designed to form a stressed part of a racing car chassis, and producing in excess of 400 horsepower.

The major components of the Ford DFV engine. The majority of engines were maintained exclusively by Cosworth Engineering until the end of 1970.

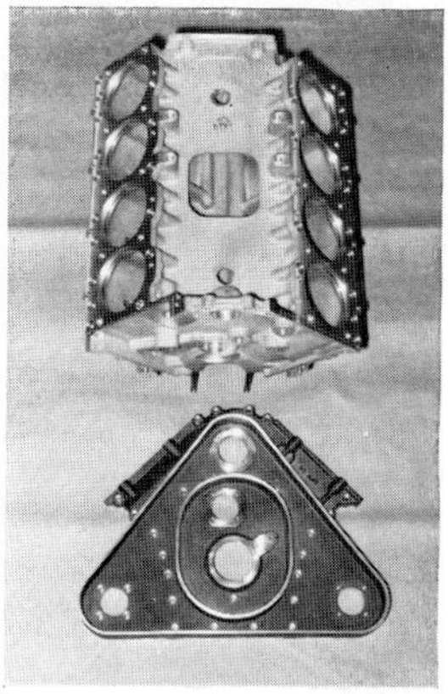

Far left: Timing gears mounted on the front end of the cylinder block. Excessive backlash was the cause of an early reliability problem.

Left: The 90-degree V8 cylinder block, cast in LM 8 WP heat-treated aluminium alloy, and the engine's front cover.

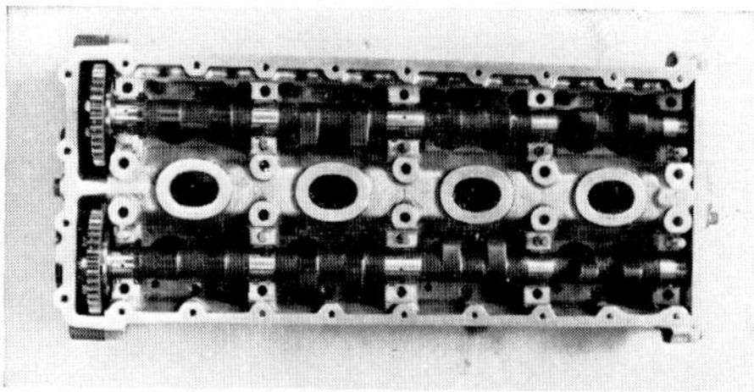

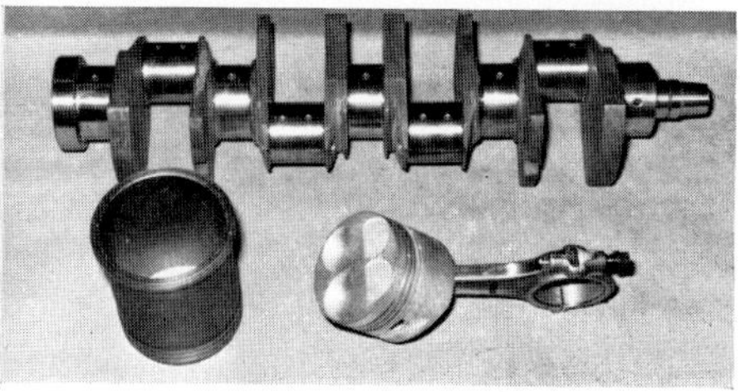

More Duckworth-designed DFV components. In clockwise order from the left: piston, connecting rod, cylinder liner and crankshaft; cylinder head; camshafts and carrier; oil pump; water pump. To overcome an early lubrication problem a larger-capacity oil pump was designed and built in time for the engine's first full season of Formula 1 racing in 1968.

Duckworth and Ford's Harley Copp examine the FVA Formula 2 engine which preceded the D

Copp, Hill and Chapman listen to a 'Duckworthism' at the DFV unveiling at Ford's London offi

Formula 1 racing, a month before DFV-day at Zandvoort. This picture taken during the early laps of the Monaco Grand Prix sums up the strength of the opposition awaiting the Ford engine's debut. Denny Hulme (Repco Brabham V8) leads Lorenzo Bandini (Ferrari V12), Jackie Stewart (2.1 litre BRM V8), John Surtees (Honda V12), Dan Gurney (Eagle-Weslake V12), Bruce McLaren (McLaren-Ford V8 modified Indianapolis engine) and Jim Clark (2 litre Lotus-Climax V8).

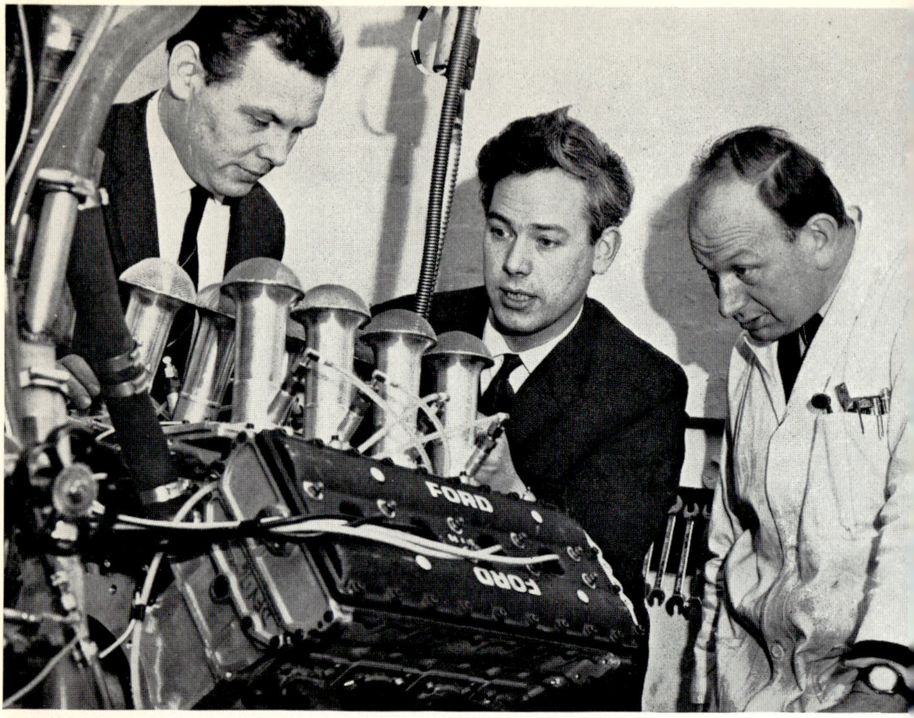

Bill Brown, Keith Duckworth and Mike Costin share a tense moment in the Cosworth test house

Jim Clark making history with a first-time victory with the Lotus-Ford 49 at Zandvoort.

Graham Hill looks pleased after his first test run with the Ford DFV-engined Lotus 49 in England.

One wheel awry, Clark scores the US GP victory which Hill had hoped to win for Lotus and Ford.

But it's all smiles as Walter Hayes is about to be the recipient of some of Jimmy's champagne!

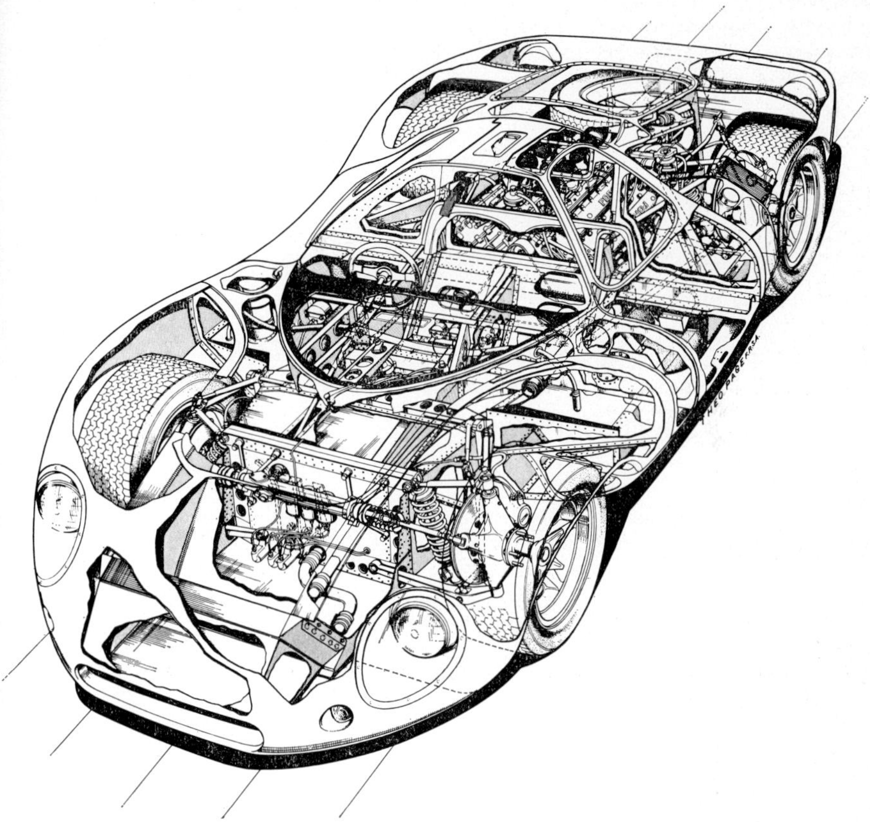

In addition to its main function as a successful Formula 1, and in 2½ litre DFW form, Tasman engine, the Ford V8 power unit was also used as the heart of Ford's experimental 3 litre P68 sports-prototype which, along with its successor, the P69, might well have been seen at Le Mans but for the switch in emphasis towards 5 litre engines for endurance racing.

The P68 on form at Brands Hatch in 1968 when Bruce McLaren challenged the works Porsche

Two months later Bruce was a surprised but happy winner of the Belgian Grand Prix with h McLaren-Ford.

Jackie Stewart's mastery of the wet at Zandvoort gave Matra their first-ever GP victory in 1968.

Another 'first' for Jo Siffert, ahead of Amon and Ickx at Brands Hatch in Rob Walker's Lotus-Ford.

Denny Hulme, McLaren Formula 1 team leader since 1968, scored his first Grand Prix success with Ford power at Monza that year.

The McLaren M7A was the first Formula 1 car after the Lotus 49 to make use of the DFV engine and like the Lotus it won its first race.

Hill v Stewart at Mexico in 1968; the battle decided the race and the championship – for Hill.

Monaco-master Graham Hill, on his way to his fifth GP victory there in the 1969 Lotus-Ford 49B.

Stewart was determined to clinch his 1969 title with a win at Monza and he succeeded – just!

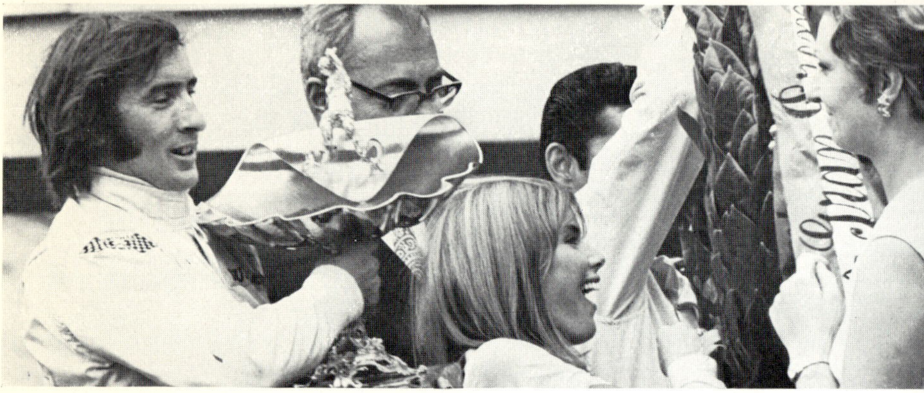

Helen Stewart is delighted as her husband is declared Italian GP winner and world champion.

They go in for large trophies in Canada. Jacky Ickx after his GP success in a Brabham-Ford.

Jochen Rindt racing his Lotus-Ford 49C to his first F1 win – the 1969 US GP.

miles from US GP winner Rindt and Man of the Race Piers Courage, whose Brabham-Ford nished second.

ack Brabham, garlanded South Africa after his th and final GP win.

Chris Amon in the March 701 – yet another DFV-engined car.

he Rindt-Brabham battle, highlight of the 1970 British GP, was to continue in the paddock.

Jackie Stewart in the Tyrrell, his third make of Ford-engined GP car, with which he was to face the Ferrari challenge with distinction.

Emerson Fittipaldi watches the bubbles at Watkins Glen. His surprise win in the 1970 US GP confirmed Jochen Rindt's world championship.

Jochen Rindt in his 1970 Lotus-Ford 72 – a final tribute to a great fighter.

initial reactions to Ford DFV power: 'I drive-a the car and I think-a Mama-Mia, this is a good-a engine. And then I get-a to 6,500 and "Bang", suddenly I have-a *two* engines.'

Clark's revs had not been as low as 6,500 since his pit stop, and with 20 laps to go he suddenly found himself elevated to fifth place as McLaren suddenly dropped out with a faulty cylinder liner, having been up in third place at one point, and Amon made a pit stop to complain of erratic handling, which was traced to a broken shock absorber mounting. By lap 56 Clark had passed Rindt and was now in fourth place, and it was clear that at the speed he was travelling he would catch both Surtees and Brabham before the end of the race to make it a Lotus one-two. Meanwhile, Baghetti's good drive had ended with a completely wrecked engine, and then there was an even bigger blow for Team Lotus when Hill's engine expired dramatically on lap 60. Graham managed to free-wheel his car as far as the pits straight and then climbed out, bitterly disappointed at having lost yet another race which he had led so convincingly.

Brabham, therefore, was the new race-leader, and Surtees was closing in on him and prepared to fight him all the way to the line. But Clark was closing in even faster, and he passed them both to take over the lead on lap 61. In his fantastic drive he had pulled back a complete lap on the people with whom he had been dicing during the early laps. It was a prodigious performance, and as Jimmy shook off the Brabham and the Honda, everyone prepared to cheer him across the finishing line. He went into the final lap, comfortably in the lead, but then came the unbelievable news that he had slowed on the other side of the course, and had been re-passed by Brabham and Surtees who had rushed into the final corner side-by-side.

The groans of disappointment at Clark's misfortune soon melted into tense excitement as all eyes turned in the direction of the *Curva Parabolica*. Brabham seemed to have had the edge half-way through the curve, but then had run wide, and Surtees had tucked inside him. There was precious little between them as they stormed out on to the finishing straight, but the white Honda was fractionally ahead and Surtees kept his foot hard down to streak across the line just a fifth of a second ahead of Brabham, while poor Clark free-wheeled over the line 23 seconds later completely out of fuel.

It was an incredible race with an incredible result, and those who watched it will remember it for many years to come.

Jack Brabham's second place at Monza had heightened the battle between the two team-mates for the 1967 world champion's crown, for Jack was now within three points of Denny's 43 with two races to go. Each driver had a burning desire to take the title, Denny for the understandable reason that he had never won it before, Jack because by winning it for the fourth time he would be one large step closer to his long-cherished ambition to equal Juan-Manuel Fangio's record of five world titles. There was not to be much give-and-take in the Repco Brabham team, therefore, during the next few weeks, and when Brabham won the non-championship International Gold Cup race at Oulton Park before setting off for the United States Grand Prix at Watkins Glen he gave further notice to Hulme that if he expected to be the next world champion he would have to work mighty hard for the privilege.

But Lotus and Ford, no longer in the running for the title, were more concerned with individual race wins, and in particular with winning the United States Grand Prix. After all, the race was taking place virtually in the Ford Motor Company's front garden, and although the Formula 1 engine was strictly a European Ford project, Walter Hayes in particular was anxious to present his colleagues in Deaborn with a race victory – or if possible a race one-two – just to prove to them how worthwhile the project had been. He decided to make the trip to Watkins Glen to see for himself how things developed during the two days of practice, and he was soon very thankful that he had done so because he was shattered by what he saw during the final hour or two of practice.

Clark and Hill had been comfortably the two fastest drivers during practice, and they seemed certain to fill the front row of the grid which at 'The Glen' is restricted to two cars. Their closest opposition seemed to be Gurney's Eagle-Weslake and Amon's Ferrari, followed by the two Repco Brabhams, but the fact that the two Lotus drivers had something in hand was apparent by what happened during the final qualification session. There was a prize of 1,000 dollars hanging on fastest practice time, and this had obviously acted as a great incentive to both Lotus drivers, who seemed to be taking it in turns to go out and beat the time which had just been set by the other. It looked finally as though Clark had won this private battle, but right at the last minute Hill had gone out and put in a real scorcher, which earned him the coveted pole position and the useful dollar cheque which went with it.

The crowd, spurred on by the public address commentator, loved every minute of it, but Walter Hayes could see his plan of a Ford one-two disappearing in a cloud of smoke. Until now he had steadfastly declined to interfere with the running of Team Lotus, but this was just too much, so he stormed up to Colin Chapman and told him that he didn't much like what he had just seen whereupon Colin suggested that perhaps he would like to tell the drivers as much himself. 'I intend to,' said Hayes. 'Have them report to me in your room at seven o'clock this evening.' And with that he took off and went back to the Glen Motor Court, which becomes the temporary home of the majority of the drivers when they are racing at Watkins Glen.

Clark and Hill duly arrived at Colin's room at the appointed hour and were reminded by Walter of the importance which Ford, as sponsors of the DFV engine, attached to a good result in the race. They were also reminded that they enjoyed equal status in the team, a condition which had been attached to the arrangement whereby Hill, at Hayes' suggestion, had been invited back from BRM to rejoin Team Lotus. This meant that they were not expected to battle with each other – least of all in practice – in order to prove a point when to do so could jeopardize the team's chances of success.

Practice had proved that they stood an excellent chance of victory in the Grand Prix, and therefore, to avoid any possibility of another 'beat me if you can' session during the race, Hayes said that he proposed that the drivers should spin a coin to decide which of them should win the race and who would be placed second, subject to both cars being ahead of the remainder at the finish. Hayes also proposed that, regardless of the finishing order, all prize money should be shared equally, and that if a similar situation arose in the Mexican Grand Prix, the finishing order at Mexico would be the reverse of that at Watkins Glen.

Clark's reaction was to say that he didn't mind, and Hill murmered that it was all right by him, and so Chapman was called upon to spin the coin and Hill had the privilege of calling. Graham called 'Heads' and won the toss, and a short while later both drivers left the room having assured Chapman and Hayes that they would be doing their best to provide them with the 'one-two' result which they wanted so badly. Later that evening they all met again at the home of lawyer and race director Cameron Argetsinger, whose pre-race party was one of the Watkins Glen traditions. Hayes found himself in a group consisting of Clark, Hill, Chapman and Brabham, all of whom were chatting animatedly about their latest toys, their private planes. Walter felt very much the odd man out as the only one without his own 'wings', and he suddenly realized that racing drivers (and in Colin's case ex-racing drivers) are mostly kids at heart, who occasionally needed fathering. It made him feel less concerned that he had chosen to be beastly to them a few hours earlier!

Keith Duckworth, however, who was sharing a room with Walter at the Glen Motor Court, was not so sure. Whereas the idea of choking off eminent racing drivers rather appealed to him, he would have no part of it because it was all predicated on the assumption that the team would win provided they behaved themselves. 'Never before,' said Duckworth in his characteristically measured delivery, 'have I heard anyone decide the result of a race nearly 24 hours before it has happened with such utter confidence.' His engineering mind would not allow him to indulge in such optimism, and he viewed the whole affair with considerable mistrust. However, he need not have worried.

The race began according to the book, with Hill taking an immediate lead from Clark and Gurney. Dan squeezed past into second place temporarily before being repassed by Jimmy, but a few laps later the Eagle's challenge expired with a broken rear suspension. Moises Solana, the Mexican driver, had been handed the spare Lotus 49 for this race, but had been allowed very little practice, and when he spun off shortly after the start of the race he found he hadn't mastered the correct cockpit drill for restarting a stalled engine, and the helping hands which eventually set him on his way again led subsequently to his disqualification.

But at the front of the race everything was going according to plan. After leading the race for 40 of the 108 laps Hill gave way to Clark, who presumably would exchange places again after a further

40 laps. But shortly after dropping back to second place Hill found considerable difficulty selecting gears. His clutch had proved troublesome from quite early on, and by this stage he was having considerable difficulty in overcoming the synchro cones in the ZF transmission. He began to lose ground to Clark, and soon found that his place was being threatened strongly by Amon in the Ferrari. On lap 65 Amon slipped past into second place, then suddenly Hill's gearbox became easier to handle again and 11 laps later he had caught and repassed the Ferrari. But Amon retaliated and went back ahead again on lap 84 when Hill had another bout of trouble. Then once again the transmission trouble seemed to cure itself and Graham was chasing Amon hard on lap 96 when the Ferrari ran its engine bearings. So, after some fairly agonizing laps, the Lotus-Fords were first and second again, although of course in the incorrect order.

Once Clark had moved into the lead he had been given pit signals showing the gap between himself and Hill, and occasionally between himself and Amon. Seeing that Graham was dropping back he had eased his pace, but this had served merely to bring Amon closer, and so after a few worrying laps he had passed the pits shrugging his shoulders in a 'What am I supposed to do now?' gesture, and of course Colin Chapman had done the only thing possible and given him the 'Go' signal.

Jimmy had responded at once, and opened up a more comfortable gap ahead of Amon, but the effect of this now had been to place the two Lotuses a long way apart. There seemed to be no question of Hill being able to close the gap again following the retirement of Amon because there were now only 12 of the race's 108 laps remaining. However, the finish was to prove closer than either Clark or Hill had anticipated.

With less than four laps to go Clark's right rear wheel suddenly assumed a very drunken angle. A retaining bolt in the rear suspension had sheared and the wheel was only prevented from falling inwards further by the radius arm, which was now resting against the cam covers. Clark reduced speed and found that he could still control the car and so he pressed on, while Hill was suddenly surprised to find that he was gaining ground quickly on his team-mate. Clark's final lap was a particularly slow one, and it allowed Hill to close to within six seconds at the finish – approximately the amount of time he had lost half-way through the race when he had slowed almost to a standstill in his fight to find a gear! Naturally, he was very disappointed, but at least Lotus and Ford had been given their 'one-two' victory. Sportingly, Clark climbed out of his car and rushed back to Hill to say 'Very sorry, Mate, but they told me to', and Hill accepted his bad luck philosophically, knowing that Chapman had made the correct decision in signalling Clark to get on with winning the race.

There was a tremendous celebration in the victor's enclosure in front of the timing tower as Clark, Hill, Chapman, Hayes and Duckworth all took their turns at mutual back-slapping and champagne-swigging. It had been a near thing, but 'ifs' and 'buts' have no place in motor racing. Results are what count, and the Ford DFV engine had shown itself during the race, as it had during practice, to be the most effective power unit in Formula 1 racing. The fact that a number of directors and other senior executives of the parent company had witnessed the drama and the success was a source of particular pleasure to Walter Hayes, who flew home to England that night carrying the vast trophy on his lap, and wearing a look of great contentment on his face. Denny Hulme was also feeling reasonably pleased with life, because he had earned four more points for his third place, whereas Jack Brabham who had finished fifth had earned only two. The title fight was not over yet, but at Mexico City it was Brabham rather than Hulme who would find the odds weighted against him.

The fact that the title fight was still unresolved was reflected in the line-up for the final round of the world championship, for both Brabham and Hulme obviously wanted to conserve their cars and were content to be back on the third row of the grid, while Clark and Amon occupied the first two places, and Gurney and Hill filled the second row.

As at Monza the race began with a confusing start, in this case because the starter waved the flag sideways instead of downwards. This caused Clark to hesitate momentarily, and he was rammed from behind by Gurney's Eagle-Weslake. One of Clark's exhaust pipes pierced the Eagle's radiator, and poor Dan was out of the race within four laps.

In all the confusion, Hill had gone straight into the lead, but Clark, whose car was unaffected by the startline incident, went through into first place on the third lap, and afterwards completely dominated the race. Hill dropped out after 18 laps with a broken drive shaft yoke, and local man Solana, who again had been given the third car, retired even earlier with a broken left front suspension.

The only real opposition to Clark's runaway victory came from Chris Amon, who put in a gallant effort with his Ferrari, and seemed assured of second place until a few laps from the end when the car became starved of fuel. Eventually Amon coaxed the Ferrari to the pits, and it looked as though he would be rewarded by fifth place, but his final lap had taken more than the prescribed maximum time and so the unlucky New Zealander failed to figure in the official results.

Brabham had done his best to keep up with Clark because he knew he had to win the race in order to take the title, but there was nothing to match the Clark-Lotus combination that day, and Jack had to be content with second place, nearly a minute and a half behind. Denny Hulme had played a canny waiting game all the way through, content to remain one place behind Brabham, and so become the new champion of the world, and one of the quietest people ever to wear the crown.

For the second year running the Repco Brabhams had taken the constructors' title, but it was on reliability more than on power. Even with the latest version of the Repco engine, with larger valves, the Brabhams had been seriously down on power, and only the outstanding roadholding of the Brabham chassis and the tenacity of its drivers had enabled them to remain competitive. Jack, of course, was well aware of this, and was hoping that some new Repco cylinder heads with two overhead camshafts for each cylinder bank would help to cut back the 60 b.h.p. deficit. But even if the revised engine did work Brabham and his team-mate had the daunting prospect for 1968 of taking on not just Team Lotus but at least two other teams which would also have the benefit of the Ford engine's 400-plus horsepower. It was a tough assignment!

8

1968: Champions of the world

THEY WERE STILL cheering Jim Clark's victory at Zandvoort when Ken Tyrrell tapped Keith Duckworth on the shoulder and said, 'How about building some for me?' Even Duckworth must have been just a little surprised at the request because at that time Tyrrell was not even involved in grand prix racing. But he had Jackie Stewart on his payroll as a Formula 2 driver, and he knew that Stewart was far from happy pedalling his heavy and unreliable Formula 1 BRM.

The ambitious and talented little Scot made it known that he was open to offers, and quite a vigorous behind-the-scenes battle developed during 1967 to secure his services for the following season. Stewart was obviously a champion in the making, and no-one knew this better than Tyrrell, who had given Jackie his big break into the motor racing limelight when he invited him to join his Formula 3 team in 1964. The two men had been together ever since in a most successful entrant-driver partnership, and Tyrrell, despite his continued denials that he had either the desire or the facilities to become involved in grand prix racing, was in fact making his first moves to do just that. He felt that if he could secure some Ford engines, he might be able to persuade Stewart to drive for him in Formula 1, and once he had the promise of Stewart's services it would be comparatively simple to obtain some chassis and not too difficult to secure the financial backing necessary to operate the team.

But by the end of the season Tyrrell's own services were in demand as a team manager – by Matra. It had been largely through his efforts that the French company had become such a power in Formula 2 during 1967, for Tyrrell's very successful team of Stewart and Ickx had scored seven outright victories in F2 races with their Matra-Fords to only two by the factory team who had used virtually identical cars. The Matra management, therefore, were most anxious to retain Tyrrell's very considerable talents for their planned entry in Formula 1 racing in 1968, particularly in view of his close ties with Jackie Stewart, who of course would be a tremendous asset to any team.

The only problem was that Matra were insistent on building not only their own chassis, but also their own V12 engine – an understandable stipulation in view of the fact that their racing effort was receiving very substantial Government backing in return for the national prestige which it was expected to win for France. Stewart, on the other hand, needed little persuasion by Tyrrell that if he intended to win the world championship he would need a Ford engine in his car. For a while it looked as though the proposed Matra-Tyrrell-Stewart deal was deadlocked, but eventually a very satisfactory compromise was reached whereby Matra Sports would build and race their own cars and engines for Jean-Pierre Beltoise, who would be backed-up later in the year by Henri Pescarolo, while Ken Tyrrell would operate a separate two-car team of Ford-engined cars under the Matra International banner for Jackie Stewart and a French back-up driver, Johnny Servoz-Gavin, both teams having the support of the Elf petroleum group.

Matra International became the first racing organization other than Team Lotus to use the Ford DFV engine when they presented a hastily prepared interim car, called the MS9, for the 1968 South African Grand Prix. It was based on one of the previous year's Formula 2 cars, the monocoque chassis of which had been cut short at the rear of the cockpit and a tubular frame added behind it to support the engine and transmission. There was such a rush to complete the car that it had arrived at Kyalami still unpainted, but Stewart worked hard with it during practice to place it on the front row of the starting grid and point the message that grand prix racing's newest team was also potentially one of the strongest.

There were several other major changes for the new season. World champion Denny Hulme had left the Brabham team to join fellow New Zealander Bruce McLaren, who was also intending to use DFV engines in a completely new car, although Hulme could only be provided with the 1967 BRM-engined car for the South African race. His place in the Brabham team had been taken by Jochen Rindt, who was feeling considerably relieved at having finally completed his three-year contract with Coopers, especially as he and Brabham were due to receive the more powerful twin-overhead-camshaft Repco engines later in the season.

The Cooper team were also about to make an engine change from the heavy Maserati to the somewhat lighter and hopefully more powerful V12 BRM, and they had signed on Ludovico Scarfiotti, Lucien Bianchi and Brian Redman as regular drivers. Jacky Ickx had joined Chris Amon to strengthen the Ferrari team, while BRM were about to draw the curtain across the unsuccessful H16 and put all their effort into their new V12-engined cars with Len Terry-designed chassis which were to be driven by Mike Spence, Pedro Rodriguez and Piers Courage. Jim Clark and Graham Hill, of course, were to continue with Team Lotus, their cars being updated to Lotus 49B specification, the main change being an increase in wheelbase, while Rob Walker had ordered a similar car for Jo Siffert, although he was having to persevere with his Cooper-Maserati for one more race.

The story of the 1968 South African Grand Prix was once again one of Lotus domination. Although Jackie Stewart led initially in his chunky-looking Matra-Ford, Jim Clark went past on the second lap and afterwards was never challenged. Stewart held second place for 28 laps before being passed by Hill in the second Lotus 49B, and the Scot retired with a blown-up engine shortly after half-distance. Clark and Hill took complete control of the race, and only one other driver, Jochen Rindt, covered the full 80 laps, Rindt revealing his enjoyment of driving a more manageable car than he had been used to by putting his Repco Brabham into some spectacular opposite-lock slides. Amon and Hulme, who finished fourth and fifth, were two laps down at the finish, and Beltoise was a further lap in arrears with his Formula 2 Matra-Ford FVA, which had been specially ballasted to conform with the Formula 1 minimum weight requirements.

By winning the South African race Jim Clark had raised his total of grand prix victories to 25, one more than the previous record by the great Juan-Manuel Fangio. But tragically this was to be the last occasion on which Jimmy was to see the chequered flag waved at him at the end of a world championship Formula 1 race, though by no means the last occasion on which he would win a race in a Ford-engined Lotus 49.

Right from its inception it had been realized that the Ford engine at some stage in its life could be used for applications beyond grand prix racing, and during 1967 the decision was taken to use it in a proposed Ford long-distance sports car, as well as in a pair of Lotus 49s which were to be prepared for the Tasman series of races in New Zealand and Australia.

In order to reduce the engine's capacity from 3 litres to the Tasman limit of 2½ litres it was necessary merely to fit different connecting rods so as to reduce the piston stroke, this quick modification being carried out on the third and fourth of the seven DFV engines produced during 1967 (the original schedule of five engines was increased to seven to compensate for two blow-ups). To assist identification, the two Tasman engines were redesignated DFW, as were all subsequent Ford engines for as long as they were fitted with Tasman rods.

There were eight races in the 1968 Tasman series, and Clark was to lead all eight and win four of them. He dropped out of the first two New Zealand races, both of which went to Chris Amon in his V6 Ferrari, but he reappeared for the third race at Christchurch with his Lotus-Ford resplendent in a new red, white and gold colour scheme, the first visible evidence of Colin Chapman's newly concluded deal with the Player's tobacco company. Henceforth, his cars would be entered by Gold Leaf Team Lotus instead of just Team Lotus, and even though race administrators at that time were inclined to be very touchy on the subject of advertising decals on racing cars there was no mistaking the message which was conveyed by Clark's car in its gay paint finish.

The change of colours also brought a change of fortune for Jimmy, who won the Christchurch race despite a strong challenge from Amon. He would also have won the final New Zealand race at Teretonga had he not left the track in the closing stages and allowed Bruce McLaren through to win with his 2½ litre V12 BRM (another reduced-capacity Formula 1 car), but he made no mistakes when the Tasman circus moved on to Australia.

Jimmy by this time had been joined by Graham Hill, and the two Lotus-Ford drivers finished first and second at both Surfer's Paradise and Warwick Farm, Sydney. Clark also went on to win the Australian Grand Prix at Sandown Park, with Hill third behind Amon, and he clinched the Tasman Championship in the final race at Longford, Tasmania, although the race itself, which was held in appallingly wet conditions, went to Piers Courage after a brave drive in his more manageable Formula 2 McLaren-Ford FVA.

Bruce McLaren, meanwhile, had flown home after his Teretonga success to oversee the final assembly of his new Ford-engined M7A Formula 1 cars, with which he was soon putting up some very spectacular lap times during private test sessions. Like the Lotus 49, the new McLaren used the engine as a stressed link between the monocoque chassis and the rear suspension and wheels, the engine being bolted directly to the cockpit bulkhead through mounting plates. Much of the basic design had been carried out by Robin Herd in close collaboration with Bruce, and the cars were superbly finished and most attractively presented in their bright orange team colour.

McLaren's methodical and painstaking development driving during the first two weeks of March paid handsome dividends, and the team arrived at Brands Hatch for the *Daily Mail* Race of Champions with the best-handling cars in the paddock. A good turn-out of cars had been promised for this first non-championship Formula 1 meeting of the year, despite the absence of the Eagle and Brabham teams, but the entry was to be considerably depleted before the end of practice. Ferrari lost one of their three cars when de Adamich crashed badly and suffered injuries which kept him out of racing for most of the year; Siffert crashed Rob Walker's new Lotus-Ford, and the damaged car was totally destroyed by a disastrous fire at Rob's garage the following evening; and neither John Surtees' Honda nor Jean Pierre Beltoise's V12 Matra were quite ready, while Surtees' substitute Formula 2 Lola was withdrawn after practice problems.

Nevertheless, it was an interesting and informative race, with Bruce McLaren leading from start to finish from his pole position on the grid, to win from Rodriguez in his V12 BRM, Hulme in the second McLaren (which had been completed only a few days before the race and was still not completely right), and Amon in his

Ferrari. Both the BRMs revealed a refreshing degree of competitiveness during this race, Mike Spence holding a firm second place for nearly half the race before losing all his oil through a detached pipe, and Pedro Rodriguez storming right through the field to within sight of the McLaren after being delayed at the start with an engine which refused to fire until long after the other cars had left the grid.

Jackie Stewart, however, found his new Matra MS10 something of a disappointment on its first outing, and had to work very hard indeed to control the car over the Brands Hatch undulations. But the Matra was merely revealing its lack of development, and both Tyrrell and Stewart were confident that given time it would become a very effective racing car. The French car differed from the Lotus and the McLaren in that although the engine was mounted similarly at the cockpit end the rear suspension was attached to a subframe which in turn was connected to the remainder of the chassis by longitudinal members running forwards beneath the engine. This meant that the rear suspension was not disturbed when engine and transmission were removed from the chassis.

Another disappointed driver at Brands Hatch was Graham Hill, who found himself the centre of a big row during practice between the race organizers and the television company which had bought the screening rights for the meeting. The TV company objected to the advertising decals being displayed on all of the cars but most of all to the display on Hill's Lotus, which included the famous Player's sailor-man on the sides of the body. Eventually a compromise was reached whereby the sailor was blacked out but the surrounding design allowed to remain – extreme pettiness on the part of the TV company. But Hill's troubles were far from over, because after a variety of mechanical problems during practice he ran only a short distance in the race before one of his drive shafts failed and tore into his oil tank. As the team were running only one car, Clark's period of overseas residence for tax purposes being not quite expired, the race was less than a successful British debut for Lotus in their new colours.

It had been hoped that Jimmy's return to Britain would take place at Brands Hatch three weeks later for the BOAC 500 endurance race, and that he would drive one of the two brand new Ford F3L sports-prototypes which had been designed by Len Bailey and built at the Weybridge, Surrey headquarters of Alan Mann, who had also been entrusted with the cars' development and race programme. But the BOAC race coincided with a Formula 2 meeting at Hockenheim, Germany where Clark was already committed to drive a works Lotus.

The Fords, which later became known as P68s, were full monocoques with very compact, wind-cheating bodywork, and weighed less than 1,500 pounds without fuel, so that with a power output of around 420 b.h.p. from their DFV engine they had a top speed potential of something approaching 200 m.p.h. with the appropriate gearing. However, Brands Hatch is a handling rather than a top speed circuit, and both cars required a lot of systematic chassis tuning during practice, much of this work falling to the very capable Bruce McLaren, one of the four drivers who had been nominated for the Fords. During the two days of training Bruce so improved the cars that he was able to set the second best practice time in the one he was to share with Denny Hulme, while Mike Spence was fifth fastest in the car he was to drive with Jochen Rindt. But the second car developed an engine fault and had to be withdrawn, and so McLaren and Spence, the two drivers with the most experience of the cars during practice, were named for the one remaining entry.

Bruce took the first driving spell in the race, but unfortunately he missed second gear in accelerating away from the start, and was half-way down the field before he could find the gear and carry on racing. He then electrified the crowd by his tremendous fight back through the field, passing all the Ford GT40s and works Porsches to claim first place inside 20 laps. The Porsche drivers, notably Siffert, fought back strongly, but although he was repassed temporarily, McLaren was soon back in front again and was still there when the first rush of pit stops and driver changes took place after two hours of racing. McLaren handed over to Spence, but a few minutes later Mike was forced to park the car on the grass opposite the Ford pit with a broken drive shaft coupling. It was a sad ending to a great effort, but there was an infinitely greater sadness to come.

A few minutes after the Ford had retired a report came through with the unbelievable news that Jim Clark had been killed at Hockenheim. The Press, naturally, had been the first to hear and after a short period of stunned numbness many of the journalists in the crowded Brands Hatch press box abandoned their interest in the exciting race before them and went about their tense and harrowing task of writing 'Jim Clark' copy. In times like these, when everyone is writing about someone they have known and admired, and whose company they have enjoyed, inter-paper rivalries do not run very deeply. Everyone helps each other with the cold facts of the driver's career, the basic information which news editors and the public demand. It is the interpretation, the slant of the story, and the personal assessments which differentiate the coverage of one newspaper or journal from that of the other.

Walter Hayes, who had been one of the first to hear the news, acted with both foresight and feeling, on the one hand seeking out every journalist he could find and quietly telling him the facts as he understood them, and on the other asking that the news should not be discussed within earshot of the public. Many thousands of families were at Brands Hatch that day having paid to enjoy a motor race. The race they were watching happened to be an unusually exciting and close-fought one between the Ford GT40s (one of which ultimately won by just 22 seconds) and the Porsches. Hayes felt that it was wrong that their enjoyment should be removed, they would learn the sad truth early enough. Accordingly he suggested that no announcement of the Hockenheim accident should be made over the public address system, and indeed none was made. Inevitably, however, there were people present with portable or car radios and the tragic news had filtered into the grandstands and enclosures by the end of the afternoon. But many thousands left the circuit after the race that afternoon still unaware of what had happened in Germany, and perhaps just a little puzzled at the lack of the customary laughter and gaiety they had observed in the pit road at the end of what had been such a magnificent race. It was a race which was to receive scant mention in the Press the next day.

A moving tribute to Jim Clark took place three weeks later at Silverstone, when the 14 drivers who took part in the *Daily Express* International Trophy race stood silently by their cars on the starting grid for one minute as a trumpeter played a lament, and many pairs of eyes were focused on a sad-faced Graham Hill, on whose shoulders rested not only the task of bolstering the morale of his understandably shaken team, but also if possible of guiding them to the world championship victory which Clark had seemed so likely to gain in 1968.

It was perhaps as well for Graham that this was a non-championship event at Silverstone, for his Lotus was to last only 12 laps before retiring with a broken fuel line. As at Brands Hatch the McLarens proved to be the pacesetters, and this time it was Denny Hulme, the fastest in practice, who went through to a clear-cut victory, with Bruce McLaren backing him up in second place. But Hulme had a few anxious moments along the way, because a stone from another car dislodged one of the lenses in his goggles, and he lost several places before he was able to adjust himself to the painful driving conditions. Chris Amon also had vision trouble, and for three laps drove without his goggles when their strap broke, but nevertheless he finished strongly in third place, while Jacky Ickx, driving a newer and lighter Ferrari, followed his team-mate across the line to finish fourth.

Once again the BRMs looked impressive for a while and Spence snatched the lead from McLaren during one of the early laps, then hounded him strongly after Hulme had passed them both. Spence had looked all set to take second place when his engine broke a timing chain a few laps from the end, and clearly the BRMs were proving a bigger threat than they had been during the first two years of the 3 litre formula. Jo Siffert was back in action with the ex-Clark Tasman Lotus, pending the delivery of Rob Walker's own replacement Lotus 49B, but a serious transmission vibration prevented him from showing his true worth, and he dropped out at half-distance. The race, therefore, was a further boost for the McLaren team's rising fortunes, although Bruce was mindful of the fact that there were several noteable absentees from Silverstone, including the Brabham and Tyrrell teams.

They had been very busy that weekend at Madrid, taking part in a Formula 2 race and using the opportunity to slip in some useful test miles on their Formula 1 cars prior to the Spanish Grand Prix, which was to take place on the Jarama circuit two weeks later. Stewart, in particular, had been hard at work testing the Matra-Ford, although he was destined not to take part in the Grand Prix. Right at the end of practice for the Formula 2 race he had slid off course into a wire-mesh fence and damaged his right wrist. At first it was thought to be only sprained, but within a few hours it had swollen badly and Stewart appeared the following day with his wrist and arm in plaster. It was to be several weeks before he was able to climb into a racing car again, and several months before he was able to drive without pain. The accident couldn't have happened at a worse time of the year, and it almost certainly cost him the world championship in 1968.

The Spanish Grand Prix was only the second occasion on which Formula 1 cars had raced at Jarama. A preliminary non-championship race had been staged there in November 1967, but it had attracted only a small entry, and as Team Lotus were the only factory team present Jim Clark and Graham Hill had simply walked away with it. This time the field was still comparatively small, but of generally high quality.

Beltoise was given Stewart's place in the Matra-Ford, the V12 Matra still being unready, and the works Repco Brabham, BRM and Lotus teams were all one car short. Brabham's new Repco four-cam engine blew up during final practice, and both BRM and Lotus suffered agonies through another motor racing tragedy. Whereas Jackie Oliver was being tipped as Jim Clark's Formula 1 replacement in Gold Leaf Team Lotus, Mike Spence had been signed as Jimmy's successor in the Lotus Indianapolis team, and had flown to America to test the car and complete the mandatory 'rookie test' for drivers new to the circuit. Mike had performed magnificently, passing his test with flying colours, but during a subsequent test run the car failed to slow for a corner, charged into a retaining wall, and ripped back a front wheel which so injured Spence that he died in hospital a few hours later. This second tragedy, so close after the death of Jim Clark, hit Colin Chapman so badly that he thought seriously of pulling out of racing for good. He had intended to fly down to Spain, where a much-modified Lotus 49B with wedge-shaped bodywork was awaiting Hill in addition to his normal car, but he felt he couldn't face another race circuit so soon, and instead sent strict instructions to his people in Spain that the new and untested car was not to leave the garage. Hill was to practise his familiar car only, and if that failed during practice if necessary he would withdraw from the race.

Jarama had been described by Hill as a 'Mickey Mouse' circuit, and by certain other drivers in even less complimentary terms. It was certainly very tight for Formula 1 cars, and its seemingly never-ending succession of slow corners and short straights put a great strain on transmissions and an extra premium on good traction. But it seemed to suit the Ferrari 312 very well, and Chris Amon put in easily the fastest time in practice to share the front row with Rodriguez and Hulme. A brand new and slightly lighter Ferrari had arrived during practice as a back-up car for Amon and Ickx, but team manager Mauro Forghieri decided to leave it on the transporter as there had been insufficient time to test it properly.

It was Rodriguez who made the best start from the front row of the grid to put his V12 BRM into the lead of a world championship race for the first time, but after 12 of the scheduled 90 laps Beltoise went past into first place with the Matra-Ford and the following lap Amon followed him through to push the BRM back to third. But then Beltoise had to make a pit stop to replace an overtightened oil union, and Amon's lead was further consolidated when Rodriguez crashed into a fence after trying desperately to keep up with the Ferrari.

Amon looked all set to chalk up his first victory in a world championship race, but as was to happen so frequently during the next three seasons bad luck robbed him of his deserved first place; in this instance a fuel pump failure stopped him after 57 laps. Ickx had dropped out early with an ignition failure, and Rindt's Repco Brabham lost all its oil pressure shortly after the start, so the race was now between Graham Hill in his Lotus and Denny Hulme in his McLaren.

Hulme had started slowly but was now hounding the Lotus hard, and clearly intent on repeating his Silverstone win. But after breathing in the Lotus-Ford's exhaust fumes for many laps Hulme began to feel unwell and dropped back in order to gulp some cleaner air. Then, feeling refreshed, he began to close the gap again only to lose the use of his all-important second gear. After that it was all over, and Hill went on to score a morale-restoring victory for the Lotus team, a quarter of a minute ahead of Hulme. They were followed home by the BRM-engined Coopers of Redman and Scarfiotti, which had proved very reliable if rather uncompetitive on their first outing, and the Matra-Ford of Beltoise, which had run faultlessly after its pit stop and set the fastest race lap.

The Frenchman had his own all-French Matra MS11 in time for the Monaco Grand Prix two weeks later, and as Jackie Stewart was still incapacitated Johnny Servoz-Gavin was able to take up his place as a member of the Matra International team. It proved to be a sensational, if rather brief debut. Having been credited with a practice time second only to that of the Monte Carlo master Graham Hill, Servoz-Gavin found himself on the front row of the 16-car grid, and with the glory of France at stake he shot straight into the lead of his first genuine Formula 1 race. The wily Hill watched the tail-wagging Matra from a safe distance, convinced that something dramatic would happen at any time, and sure enough on his fourth lap Servoz-Gavin snicked the wall of the chicane leading on to the harbour front, and limped slowly back to the pits to retire, to be followed a few minutes later by Beltoise, who had had a similar experience with the V12 Matra. After Servoz-Gavin's exit it was Hill all the way, driving his new wedge-shaped Lotus 49B in impeccable fashion to score his second GP victory in two weeks and his fourth at Monaco. His new team-mate, Jackie Oliver, had a miserably brief race, tangling with Bruce McLaren's already disabled car on the opening lap, but the star of the race was another relative stranger to Formula 1 racing, Richard Attwood, who had been signed on to replace Mike Spence in the BRM team. He took over second place when Surtees retired his Honda with a broken transmission (a repeat of his trouble at Jarama), and harried Hill all the way to the finish, a new record time for his 80th and final lap taking him to within 2·2 seconds of the Lotus at the finish. It was a magnificent drive which, unfortunately, Attwood was unable to repeat in his subsequent races with the team.

Once again the Cooper-BRMs ran reliably, third and fourth places this time going to Lucien Bianchi and Ludovico Scarfiotti, and the only other finisher was Denny Hulme, who lost six laps having a broken drive shaft replaced on his McLaren. Curiously, all the 11 retirements occurred during the first 17 laps, the majority being suspension failures or other damage caused by hitting immoveable hard objects, although Gurney's Eagle-Weslake retired with a broken engine. Hill was now comfortably in the lead for the world championship with 24 points against the 10 of Hulme, but of course there was still a long way to go.

The next visiting place was the Spa-Francorchamps circuit for the Belgian Grand Prix, interest in which was heightened by the fact that Jackie Stewart was making his return to racing, despite having his right wrist still encased in a special lightweight plastic cast. Also, Ferrari, who had given the Monaco race a miss, were back in force and they carried a 'secret weapon' in their transporter, as indeed did Jack Brabham in his.

As engine power had soared over the 400 b.h.p. mark, and tyres had grown progressively wider, the problem of adhesion had become very considerable in grand prix racing. Horsepower is essential for winning races, but it is an embarrassment rather than an asset if it cannot be transmitted through to the road. A similar problem existed in large-capacity sports car racing, and the American driver-designer Jim Hall, a most resourceful and deep-thinking young man, had found a most effective method of overcoming it on his Chaparrals. He had mounted a wing over the rear of the body, the blade being angled and profiled so that it would work like an aircraft wing in reverse and provide a downward thrust rather than a lift, or in aerodynamic terms a negative rather than a positive lift. It seemed only a matter of time before similar high-mounted wings would sprout from grand prix cars, and to Ferrari must go the recognition for being the first (just) to fit them in public, although Lotus had been the first to fit small wing extensions to their nose cones with the 49B seen at Monaco.

Chris Amon was able to knock several seconds off his previous best lap time after he had been provided with a wing towards the end of the first day's practice, and as the following day was terribly wet he secured pole position very comfortably. Ickx decided not to use a wing on the other Ferrari, but Amon's time had encouraged several others to rush around with tin snips and spanners in an effort to create more downthrust on their cars. Before the race the Repco Brabhams had appeared with small wings over their gearboxes in addition to the nose tabs which had been fitted back in England, and the two works McLarens had 'tea trays' of different sizes extending from the rear of their engine covers. Surtees also had found room for a small tray between the bunched exhausts of his Honda's engine.

The aerodynamic aids certainly seemed to work in the early stages of the race as Amon streaked off into an immediate lead, hotly pursued by Surtees, who squeezed past in the Honda on the second lap. Amon then tucked in right behind, which proved his undoing because shortly afterwards a stone thrown up by one of the Honda's wheels pierced the Ferrari's radiator, and Chris had to abandon the race as the last drop of water ran out. Surtees built up a comfortable lead over the first ten laps, but suddenly he had a big fright as his rear suspension collapsed, and after bringing the wildly sliding car under control he drove slowly back to his pit.

Hulme, who had been dicing hard with Stewart, now found himself in first place, and the battle between the McLaren and the Matra see-sawed for several laps until Hulme broke another drive shaft and retired after 18 of the 28 laps. His retirement put Stewart into a commanding lead over the eight other drivers still in the race; both the Repco Brabhams had retired early with engine problems and Hill's Lotus had broken a drive shaft, so the Scot looked set for nine championship points with which to celebrate his return to the cockpit. But with just over a lap to go the Matra ran out of fuel and Stewart coasted to his pit. Bruce McLaren and Pedro Rodriguez, who had been dicing for second place, went past, now dicing for the lead, although they had been so preoccupied that both of them failed to notice Stewart's car stationary at the pits. Jackie Ickx, whose Ferrari had been misfiring for much of the race, also went past before Stewart was able to get away again, so he was now fourth and feeling very disappointed. Bruce McLaren, on the other hand, was bewildered. He had pulled out 12 seconds on Rodriguez during the final lap, and he was quite taken aback by the waving and cheering which greeted what he assumed to be second place as he acknowledged the enthusiastically waved flag. It was only after he had driven back into the paddock and someone told him that he was wanted on the winner's rostrum, that he realized he had won his first Formula 1 grand prix for six years! Ford-engined cars had taken three of the first six places, although Jackie Oliver's smooth drive into fifth place with a new Lotus 49B was halted on his final lap with a drive shaft failure, his place only being retained because Bianchi's Cooper-BRM, the next car to finish, completed only 26 of the 28 laps.

It was now a full year since the DFV engine's debut at Zandvoort (the order of the racing calendar had been changed slightly for 1968), and already it had achieved an impressive record of success. Of the 13 world championship races in which it had been entered, it had powered the winning car on eight occasions, in addition to winning three non-championship Formula 1 races and, in DFW form, four Tasman Championship events. It was a formidable performance for such a new engine, and already there were signs of some deterioration in the quality of the opposition, of which only Ferrari seemed to be capable of offering any substantial sustained challenge.

The latest four-valve-per-cylinder V12 engine seemed to be producing a lot of power, even if it was confined to a comparatively narrow rev band, and the Ferrari team might well have figured more prominently in the early 1968 results but for a series of minor problems. The Eagle-Weslake, on the other hand, which had seemed likely to offer the greatest threat to Ford engine success, had failed to live up to its earlier promise. That the engine was capable of producing adequate power there had been little doubt, but various manufacturing problems had resulted in a poor reliability record, and the team's problems were compounded by the fact that a number of engine components had proved not to be interchangeable. This meant that in the event of a component failure in one engine it was often impossible to utilize the equivalent component from another engine in order to rebuild it. As a result, Dan Gurney was finding himself desperately short of power units, even for just one car, and it had been particularly galling for him to have to miss the Belgian Grand prix, the scene of his big success in 1967.

John Surtees had persevered as only he knew how with the V12 Honda which, even in the much lighter Lola-inspired chassis, still added up to a grossly overweight grand prix car, while a transmission design weakness had added considerably to Surtees' problems. The Cooper team, meanwhile, had been pleased to accept the BRM engine as a replacement for their earlier Maseratis, but it was of little comfort to them that with similar power at their disposal their cars were no match for the works BRMs, even though they might just have had the edge on reliability. The team also had a most unhappy Spa weekend, for news filtered through from Germany on the eve of the race that Ludovico Scarfiotti had been fatally injured whilst practising a Porsche for the Rossfeld hill-climb, and in the Grand Prix a front suspension failure had sent Brian Redman off course at high speed, resulting in a serious arm injury which was to keep him out of racing for the remainder of the season.

It was too early in the life of the Matra V12 engine to make any firm assessment of its potential, but certainly it had achieved little so far to suggest that it would pose any serious threat to the Ford. In fairness, however, the Matra had been racing at Spa with the ultra-long exhaust pipes which had been fitted for its Monaco debut in an effort to provide maximum middle-range torque at the expense of top-end power; these were quite unsuitable for the very fast Belgian circuit, but a strike in France had prevented more suitable exhausts from being made in time.

But the big surprise on the engine front had been the poor reliability of the four-cam Repco V8. Although few if any people expected that the revised engine would be able to match the Ford DFV on power, it was assumed that the excellent reliability record of the earlier Repco would be maintained. But Brabham and Rindt had suffered a whole variety of problems, ranging from dropped valve inserts through sticking throttle slides to at least one major blow-up, and no Repco Brabham had finished a race since the South African Grand Prix five months earlier.

However, for the Dutch Grand Prix there were to be three works Repco Brabhams, so that mathematically at least the chances of success were brighter. Dan Gurney, still without an engine for his Eagle, was offered the third works car, and Silvio Moser was also present with his ex-works BT20 to swell the Repco Brabham contingent. All four cars qualified for the race, and with Rindt in the centre of the front row between Amon and Hill, and Brabham immediately behind and alongside Stewart, it looked more promising for the Byfleet team. Also, the weather at Zandvoort on race day recalled the monsoon-like Canadian Grand Prix of the previous August, when the Repco Brabhams had finished first and second. But it was to be very different this time.

Dunlop had developed two particularly effective types of wet-weather tyre, and Stewart and Beltoise made the best possible use of them. Taking over the lead from Hill on the fourth lap, Stewart, still driving in considerable pain from his cracked wrist, took complete command of the race and held it right through to the 90th and final lap, his Matra-Ford finishing just over a minute and a half ahead of Beltoise in the V12-engined car. The Frenchman, who was the only other driver to complete the full distance, had a particularly busy race, for he spun off at least twice, and survived mammoth slides on several other occasions before finally acknowledging the chequered flag. Matra, of course, were delighted with his performance with their 12-cylinder car, even though the appalling conditions had contributed considerably to its high placing.

Rodriguez contrived to keep his BRM on the track during another typically brave drive and was rewarded with third place, but nearly everyone else spun off at least once, and in some cases several times. Practically all the retirements – and they included all three works Repco Brabhams – were caused by sliding off the track, water-logged ignition systems, or by the impossibility of controlling the car on the streaming wet track, a factor which made Stewart's victory – his first in a grand prix for two years and his first of many for Matra – all the more meritorious. But if anyone deserved an accolade that day it was the design staff of Dunlop!

The weather was equally depressing for the French Grand Prix at Rouen two weeks later, but there the misery was to be compounded by yet another motor racing tragedy, in this case one which so easily could have been avoided. A few days before the race John Surtees was to find out that he did not, as he had been led to believe, have control over the operation of Honda's racing effort. The Japanese company had produced a completely new car, the RA 302, with an air-cooled V8 engine, and they asked that the car be taken to Rouen. Surtees complied with their wishes, but with a certain degree of reluctance, because he felt that it was pointless to expose the car before it was raceworthy; he would not consider driving it in the race and therefore it would be a waste of time and a division of effort to attempt to test the car during the practice sessions. Like most entrants and drivers, Surtees was firmly of the opinion that the early tests of a completely new design should take place in the anonymity of a disused circuit, where its strengths and weaknesses could be evaluated privately and without interruption.

But Honda had other ideas for their new car. They were engaged in a marketing campaign for their road cars in France, and so they decided to wave the corporate flag by entering the car in the race. Surtees was furious, but the company was adamant, and they named the likeable Jo Schlesser, who had associations with Honda in France, as its driver.

Schlesser was a successful and experienced long-distance driver, but he had had very limited experience of grand prix cars, and consequently he felt highly honoured at being given the chance to race in his own country's Grand Prix. He worked extremely hard throughout practice attempting to learn the characteristics of what was clearly a very odd-handling car, the oddness being compounded by the fact that the driving position was unusually far forward and therefore gave the driver very little feel of what was happening back at the driven wheels.

To put such an under-developed and unorthodox car in such relatively inexperienced hands on a more than usually dangerous circuit was a gross error of judgment, and it needed only the additional hazard of rain immediately before the start to provide the final ingredient for a tragedy. It occurred on only the second flying lap, when the Honda failed to take one of the very fast and very tricky bends on the downhill section connecting the pit area with the lowest point on the circuit. The car went straight on into the bank on the outside of the bend, and being made largely of magnesium it exploded on impact and burnt furiously; poor Schlesser must have died instantly. It was barely possible for the race to continue, for the burning wreckage covered a considerable portion of the track, but somehow the other cars were threaded past the accident, and the Grand Prix carried on, though few of the drivers had much stomach for it.

Within a few laps the rain had intensified, and car after car called at the pits for a tyre change when it became obvious that the track would remain wet throughout the race. The Ferrari team had decided to play it safe, equipping Ickx's car with 'wet' tyres and Amon's with tyres with a 'dry' compound and tread pattern, a wise decision based on the knowledge that Amon was always at his best in the dry, whereas Ickx could be expected to excel in the wet. How right they were.

The young Belgian led right from the start, and was only headed very briefly on one of the 60 laps when Rodriguez nosed his BRM ahead for a short distance during one of the heavy rain storms. Both men drove valiantly, although the Mexican was to run short of gears and retire before the finish, allowing Surtees through to second place. He finished almost two minutes behind Ickx, having been one of the many to make a pit stop on the way.

Jackie Stewart found his Matra-Ford a handful in the wet this time, having started the race on 'intermediate' tyres, and although he went well after switching to 'wets' he had lost too much time during the earlier laps and was only able to finish third, one place in front of Vic Elford. Elford was having his first drive in a grand prix car as Brian Redman's substitute in the Cooper-BRM team, and he did magnificently to run through non-stop on 'dry' tyres, no doubt benefiting from his immense experience as an international rally driver! Elford's team-mate for this race was Johnny Servoz-Gavin, for whom no Matra-Ford had been available since Monaco, but the Frenchman left the course, fortunately without injury.

But the luckiest man at Rouen that weekend was Jackie Oliver, who during one of the practice sessions suddenly found his Lotus spinning out of control on the straight before the pits. The car must still have been travelling at 100 m.p.h. when it hit a barrier backwards and broke up into several pieces, and it was a miracle that Oliver was able to climb out of the cockpit section shaken, but otherwise unhurt. The cause of the accident was a mystery, but the assumption was that the Lotus, which had been travelling close behind two other cars, suddenly became unstabilized when Oliver pulled out of their slipstream to pass them. The Lotuses were making their first appearance with wings (mounted on the rear suspension rather than on the main structure of the car), and it was thought that a sudden change in air turbulence on one side of the wing had thrown the car out of line. At that time, practical knowledge of wings on racing cars was still very limited, and most of the developments by the various teams had been carried out on a 'hit or miss' basis, or on the advice of outside experts, whose fund of knowledge had been accumulated through other applications. Consequently there was still much learning to be done, and already one or two people were beginning to wonder whether the huge airfoils were such a desirable thing after all.

However, it had been established firmly that when properly mounted and adjusted they improved cornering performance and therefore helped to reduce lap times on most circuits. There was every encouragement to use them, therefore, for as long as they were admitted under the regulations, and during the two weeks between the French and British Grands Prix they had grown considerably both in number and size. However, the antics of John Surtees' Honda after his wing snapped off during the Brands

Hatch race (due to being too rigidly mounted) was further evidence of the potential danger of wings in a failure situation, as indeed was the sight of the blade itself spinning away towards the side of the track, but mercifully not into the crowded enclosures.

After their Rouen setback, Gold Leaf Team Lotus seemed to be right back on top form for the British Grand Prix, with Hill and Oliver making the two fastest practice times to share the front row with Amon and his Ferrari, while Siffert was on the inside of the second row with Rob Walker's brand new replacement Lotus 49B to provide further reason for Colin Chapman's good humour before the race.

For the first 26 of the 80 laps the race went in copybook style for the Lotus team, Oliver making the best start and leading initially, but giving way to Hill on the fourth lap, the two of them dictating the race confidently, backed up by Siffert, who was doing well to hold off a strong challenge by Amon for third place. But suddenly, Hill's car broke a universal joint and after a few anxious moments Graham slowed down to let Oliver through into a fairly comfortable lead while he made his way back to his pit with one rear wheel at a very drunken angle.

Oliver was driving very coolly on a circuit he has always enjoyed, and it looked as though he might win his first grand prix there until he ran a bearing and dropped out on the 43rd lap. So now Siffert was in first place with the third Lotus 49B, and despite a tremendous challenge by Amon in the closing stages he kept his head and crossed the line the winner by 4·4 seconds – yet another close finish to a British Formula 1 race. It was Rob Walker's first triumph since 1961, when Stirling Moss was his regular driver, and of course he was delighted, particularly as his car was still being built a few days earlier, and in fact was finished off in the Brands Hatch paddock on the first day of practice! Jacky Ickx supported Amon well by finishing third ahead of Denny Hulme, and somehow John Surtees contrived to tame the oversteering antics of his wingless Honda to take a gallant fifth place ahead of Jackie Stewart, whose wrist was particularly painful for this race, Brands Hatch being one of the toughest of the grand prix circuits on both car and driver.

But the toughest conditions of the 1968 season undoubtedly were encountered at the Nurburgring, where persistent fog completely disrupted the practice days before the German Grand

Prix and for a considerable time seemed certain to cause the race to be cancelled. After lengthy deliberations between drivers and organizers the GP took place in appallingly difficult conditions, the fog over the high plateau incorporating the start-finish area being so thick at times that visibility was down to about 50 yards. It is in conditions like these that racing drivers really show their worth, and Stewart put up an epic performance which was to earn him the award of 'Driver of the Year' by the Guild of Motoring Writers.

Ickx and Amon had put their Ferraris on the front row of the grid during the one reasonable period of practice, with Rindt's Repco Brabham on their left, whereas Stewart was back on the third row behind Hill's Lotus and Elford's Cooper-BRM. But Jackie made a brilliant start, and aided by Ickx's slow getaway from pole position, was able to slip quickly through into third place behind Hill and Amon. He was determined to seize the lead at the first opportunity so as to get free of the wheel spray of other cars, which was adding to the fog hazard, and after challenging several times he found a way past the Ferrari before Adenau, and then set off to close the gap on Hill's Lotus, which he passed on the exit from the smaller of the two Karussels.

Once in the lead he began to drive like someone who carried radar in his cockpit, and by the time he emerged through the fog into the pit straight he had a lead of eight seconds, which he proceeded to extend throughout the race at the rate of about 10 seconds every lap.

Although he was unable to live with Stewart that day, Hill was more than a match for everyone else, only Amon being able to offer him a close challenge. But with two and a half laps to go Amon had trouble with his differential and the Ferrari slid gently off course into a shallow ditch, where it remained until the end of the race. A minute or two later Hill spun his car in the middle of the track half-way round a blind corner and had a worrying few moments during which he expected Amon's Ferrari to arrive at any time, not having seen it disappear. But Graham, feeling very relieved, managed to get his car pointing in the right direction again several seconds before the new third-placed man, Jochen Rindt, arrived on the spot, and during the remaining two laps he kept the intermittently misfiring Repco Brabham at bay to claim six valuable world championship points.

But Stewart had taken the flag more than four minutes earlier, and was out of the car long before the Lotus appeared on the scene. Both he and Ken Tyrrell were feeling very cold and wet but nevertheless very elated, because not only had it been a drive which would be remembered for many years, it had put some added spice into the world championship battle. Graham Hill still headed the table with 30 points, but Stewart was now only four points behind with four races still to go.

The position was unchanged after the Italian Grand Prix, where all but six of the 20 starters dropped out and Denny Hulme hit top form with his McLaren-Ford and swept through to a clear-cut victory almost half a minute ahead of a closely battling Johnny Servoz-Gavin (back in a Matra-Ford again) and Jacky Ickx. It was an unusually large margin for the slipstreaming Monza circuit, but Hulme was the sole survivor of a quartet comprising Stewart, Siffert, McLaren and himself who had contested the lead until well after half-distance. Servoz-Gavin had been entered in this race in the hope that he could rob Stewart's championship rivals of a point or two, and he did a very useful job in beating Ickx across the line by a fifth of a second, though aided very considerably by a vapour lock in the Ferrari engine which slowed Ickx during the vital rush up to the finishing line.

For Ferrari it was a most disappointing race, after they had entered three cars in the hope of repeating their 1966 success, for Derek Bell's first Formula 1 drive ended with fuel starvation after only five laps, and Chris Amon had a most alarming accident four laps later when he slid on oil and left the track backwards at very high speed, causing John Surtees to hit the bank while taking avoiding action in the Honda. Fortunately, neither driver was hurt.

One of the most interesting aspects of Hulme's victory was that he had backed a hunch that a winged car's extra drag down the long Monza straights would more than cancel out its advantage over a wingless car through the corners, and it was not without significance that of the four cars contesting the lead for much of the race only Stewart's was equipped with a wing. On the other hand, other teams had so much confidence in the advantages of airfoils at Monza that they arrived at the circuit with their most sophisticated efforts to date. The Repco Brabhams were visibly the most spectacular in practice with tall wings at the front as well as the rear (although the front wings were discarded before the race), but to Ferrari went the top marks for technical ingenuity. Their wings were made adjustable so as to minimize drag on the straights, the system being controlled partly automatically and partly by the driver. The wings were inclined automatically whenever the cars were in first, second or third gears, but ran almost horizontally in fourth and fifth. They also dipped whenever the brake pedal was depressed, and there was an overriding switch which enabled the driver to incline his wing at will.

But despite Hulme's victory at Monza, the wings were as popular as ever again at St. Jovite for the Canadian Grand Prix, which was the first race to be contested entirely by winged Formula 1 cars, Brabham and Rindt deciding to race with wings at both ends of their cars having convinced themselves that a blade of glass-fibre a foot or two in front of their faces was not so inhibiting after all.

For the first time Matra were running a two-car team for Pescarolo as well as Beltoise, but to offset the newcomer a previously familiar V12 was missing. Dan Gurney sadly had decided not to persevere with his Eagle Formula 1 effort and had accepted Bruce McLaren's invitation to join the McLaren team for the remaining three 1968 races. The engine problems which had dogged the Eagle for so long, he decided, were insoluble without the expenditure of a lot more money on top of the very considerable sum spent already, and so with a certain amount of reluctance Dan elected to confine his Eagle activities to American racing at Indianapolis and other USAC circuits, and to revert to being just a driver in Formula 1. The Eagle-Weslake had appeared only five times in 1968 world championship races, and had retired on four occasions, usually very early in the race. Only in the treacherous German Grand Prix did the car run reliably, but even then the unfortunate Californian suffered a puncture and dropped back to ninth place at the finish. It was a sad ending for what potentially had been one of Formula 1 racing's most promising efforts. The workmanship which had been put into both car and engine was impressive, and the design concept had been good. The ultimate failure, perhaps, was the result of too much being attempted too quickly. There was an abundance of ambition, enthusiasm and dedication by both Anglo American Racers and Weslake Engineering, but both companies were a little short of first-hand experience of the complexities of building and maintaining a complete grand prix team. That relations between the two companies became strained as the magnitude of their problems increased served merely to aggravate an already difficult situation, and by the time Gurney and his staff moved away from the Weslake premises at Rye into their own factory near Ashford the writing was already on the wall for the Anglo American Eagles.

The St. Jovite circuit, some 80 miles north of Montreal, is notoriously hard on cars, due partly to the tortuous nature of the circuit and partly to its somewhat ripply surface, which sets up severe vibrations through the chassis, particularly under braking. The two days of practice, therefore, brought many problems for most teams, although for once everything seemed to go well for Rindt, who secured pole position with his Repco Brabham with a time later equalled by Amon in his Ferrari. But the Italian team suffered a serious setback during practice when Ickx crashed badly after his throttle had stuck open, and suffered a fractured leg. Amazingly, he was able to leave hospital just in time to watch the race from a seat on the pit wall, but his slim chances of securing the world championship had now disappeared. Hill had also been involved in an accident with his Lotus at almost the same spot, but the damage to his car was relatively slight and he was able to take his place on the second row of the starting grid, alongside Gurney and behind Siffert.

But the grid line-up was to bear little resemblance to the race result. Amon went into first place right from the start, followed by Siffert and Rindt, and although he lost the use of his clutch after a few laps he steadily built up a useful lead, so that when Siffert stopped with a loss of oil and Rindt with an overheating engine before half-distance the Ferrari was so far ahead that no-one stood a chance of catching it. But once again Amon's luck was out, and after completing 72 of 90 laps the Ferrari's gearbox succumbed under the strain of so many clutchless changes. Denny Hulme now inherited the lead, and remained there to the end to be followed home by Bruce McLaren, who eased up considerably during the closing laps to finish a lap behind; McLaren's second place had been made secure a few laps from the end when Rodriguez, running behind him, made a quick pit stop with a badly smoking

BRM. It had been a great day for the McLaren team, marred only by Gurney's misfortune in collecting a stone through his radiator, when leading the three-car team early in the race. A similar stone had hit, but not quite punctured Hulme's radiator, which went to prove that you need your share of good fortune in order to win races.

It was during this race that Graham Hill had his famous and hilarious 'jelly ride'. Half-way though the Grand Prix, after running second to Amon, Hill had brought his Lotus 49B into his pit complaining that he seemed to be sitting a little closer to the ground on every lap. The Lotus mechanics jacked up the front of the car and were shattered to find that the rear part stayed still, then they jacked up the rear and found that the front half failed to move. Sure enough, the Lotus was sagging in the middle, and a quick inspection revealed that the engine mounting bolts had worked loose, probably as a result of Hill's practice accident. The bolts were tightened as quickly as possible, and Hill was sent back into the race, but after taking things very steadily, because the handling was still rather erratic, he was more than a little surprised and considerably relieved to finish the race in fourth place with the front half of the Lotus still attached to the back.

Although Graham had added three valuable points to his championship score, whereas Stewart had managed only one for sixth place, having had to stop early in the race to replace a bent front suspension arm, he had now been joined at the top of the table by Hulme, whose two successive GP victories had put him right back into the title hunt with 33 points, six more than Stewart.

Jackie realized that he would have to win the United States Grand Prix at Watkins Glen to retain any reasonable chance of taking the title, but he had a rude shock during practice when Gold Leaf Team Lotus 'guest driver' Mario Andretti, making his first visit to the circuit, pipped him by seven-hundredths of a second for pole position on the grid, while Graham Hill was a bare hundredth of a second behind on the second row, alongside Amon.

But it was to be Stewart all the way in the race, using his spare Matra-Ford after his regular car had been badly scorched a few days before the race by a fire in the Watkins Glen technical centre. Jackie was chased all the way by Graham Hill, after Andretti had dropped back from second place with loose bodywork, but he crossed the line a quarter of a minute ahead after easing up considerably during the last few laps. Andretti had been forced to retire with clutch trouble following his earlier stop to have his nose cone taped up, and third place was held for most of the race by Gurney, although his McLaren had been involved in a titanic struggle with Surtees' Honda for much of the time. Only when the Honda developed a misfire was Gurney able to shake off the determined Surtees, but just when his third place seemed secure the McLaren developed a slow puncture and simultaneously ran short of fuel. Poor Gurney was reduced to a crawl on his final lap, and as he crept towards the chequered flag, his engine spluttering for lack of fuel, Surtees found sufficient power from his own sick engine to close the gap and beat him across the line in the final few yards.

By now only Hill and Stewart, with 39 and 36 points respectively, stood much chance of taking the title as Hulme had been unable to add to his 33 in what had been for him a disastrous race. Having been fifth fastest in practice, Hulme had worked his way up into third place behind Stewart and Hill in 15 laps, only to slide off course and damage a brake pipe. He had to make two stops, one to plug the oil leak, the second to have a replacement pipe fitted, but gradually he worked his way back up into seventh place only to run into differential trouble and slide off backwards opposite the pits, badly damaging his car when it careered sideways into a small bank. Hulme climbed out virtually unhurt, but the car was so badly damaged that a replacement had to be flown out for the final grand prix at Mexico City.

This was the race in which everything seemed to go wrong for Stewart. It started with a drive shaft failure in practice, which tore up the Matra's rear suspension and wing, and gave the Tyrrell mechanics a hectic time repairing the car in time for the race. Understandably, Stewart was back on the fourth row of the grid, but within a few laps he was dicing for the lead with the Lotuses of Siffert and Hill. Siffert had been fastest of all in practice, and was again driving like someone inspired as the leading trio changed places time after time. But suddenly he dropped back and stopped at his pit, losing two valuable laps while a broken throttle linkage was repaired.

Siffert's trouble had left Hill and Stewart out on their own, and by now the title fight was strictly between the two of them because Hulme had retired dramatically in another alarming moment right in front of the pits. This time one of his suspension units had given way as he entered the pit straight, and the car had swerved into the barrier and gradually ground its way to a stop, Hulme having to hop out rapidly as it burst into flames.

Everything was at stake in the battle for the lead, for whoever won the race would also be champion; if Hill won and Stewart was second Graham would win the title by 48 points to 42, but if Stewart beat Hill across the finishing line they would both end up with 45 points and Jackie would take the title on the strength of his greater number of race wins. The Grand Prix was over 65 laps, and with only 15 to go it was still either driver's race, and title, but suddenly Stewart's Matra developed a misfire and started to drop back. He was passed first by McLaren's M7A, then almost simultaneously by Oliver's Lotus 49B and Rodriguez's BRM which had been dicing hard for fourth place and were now fighting for third, then by Bonnier in the second works Honda (Surtees had long since retired with a vapour lock), and eventually by Siffert, who had been driving harder than ever after his pit stop in the hope of getting back on to the leader board and was now rewarded by sixth place and a new lap record.

With the pressure removed, of course, Hill was able to ease up for the first time in the race, and he even had time to assist Oliver by allowing him to unlap himself and then shielding him for a lap or two from the tenacious Rodriguez. For Gold Leaf Team Lotus it was the perfect ending to what had been a desperately sad season. Graham Hill, at 39 years of age, had won the Mexican Grand Prix and with it the 1968 championship. It was the second world title for the driver with the sauciest wink in motor racing, and with it went a third-time victory in the constructors' championship for Lotus, whose earlier titles had been won for them by the sadly missed Jim Clark. But this was the first time that both world titles had been linked with the Ford DFV engine, which in less than two racing seasons had become the most powerful force in Formula 1 and had already more than fulfilled its original purpose of assuring Britain's place at the forefront of international grand prix racing.

9

1969: Entente Cordiale

IN 1968, APART from being linked with both drivers' and constructors' world championships, the DFV engine had powered the winner of all but one of the 12 qualifying races and had also been fitted to the second-place car on seven occasions. Of the 11 victories, five had been scored with Lotus cars and three each with McLarens and Matras, so that no one could deny that Cosworth Engineering and the Ford Motor Company had served the best interests of motor racing by making the engine available to more than one team. Nevertheless, it was fitting that Gold Leaf Team Lotus should have been the first racing organization to win a championship with DFV power having been the first to use it, as a direct result of Colin Chapman's persistent and tireless efforts to acquire a competitive 3 litre engine.

But Cosworth were to become busier (and more successful) than ever in 1969, for a fourth team had been drawn into the DFV-user's club. Jack Brabham, who had stood by his Repco associates in Australia through thick and thin, felt that maybe things had been just a little too thin in 1968 (he had scored a bare two championship points all year and Rindt had accumulated only eight) and that it was high time he turned to Ford power. He anticipated that this would involve a completely new design of car and Ron Tauranac set about producing his first Formula 1 monocoque chassis to accommodate the Ford engine. However, tests carried out on an interim car, a 1968 spaceframe BT26 with a revised rear end to suit the new power unit, proved so successful that it was decided to build three replicas and keep the monocoque under wraps until such time as the interim cars proved uncompetitive. In fact the monocoque was not to appear in public until the beginning of 1970, when it became known as the BT33.

Rindt, meanwhile, had become disillusioned by his unsuccessful year with the team and had accepted an invitation to join Hill at Lotus, his place in the Brabham team being taken by Ickx. Enzo Ferrari seemed to be in no hurry to name a replacement for the Belgian, and Amon was told that for the time being at least he would have to carry the Ferrari challenge single-handed, which could scarcely have been a comfort for him.

But at least Ferrari were still involved in grand prix racing, unlike three other factory teams which had been a part of the Formula 1 scene the previous year. The Cooper group of companies had undergone a considerable financial reorganization, and the new management stipulated that substantial outside sponsorship would have to be obtained if the grand prix team was to continue for a further season. On the results of 1968, however, it was virtually impossible to find the necessary backing, and so the team which had done so much ten years earlier to put Britain and Jack Brabham on top in Formula 1 finally had to bow out. It was an inevitable decision after several rather uncompetitive years, but nevertheless it was universally regretted.

Honda, on the other hand, had been relative newcomers to motor racing, having made their debut towards the end of the 1½ litre formula. They had entered motor racing with the firm expectation that they would be able to match on four wheels the dominance they had enjoyed for so many years on two, but they made several fundamental errors of design as well as racing policy, and not even Surtees' best efforts could remove the tarnish from the Honda image which had been brought about by the Rouen tragedy of the previous July. It was rumoured for a time that Bonnier, who had driven the second works car in the final 1968 race, would run a works car in 1969, but no one was very surprised when Honda decided to abandon GP racing completely.

The Japanese team's withdrawal rendered Surtees 'open to offers', and eventually he accepted an invitation to lead a reorganized BRM team, with Oliver as his number-two, a move which some uncharitable observers suggested would transfer him from the frying pan into the fire. But despite BRM's poor 1968 results, both in respect of performance and reliability, there were considerable hopes for the latest four-valves-per-cylinder version of the V12 engine, and for the lighter-weight P138 chassis which had been built to carry it in place of the earlier P126 and P133 models. Significantly, however, Surtees did not mince his words at the press conference at which his new assignment was announced, and there was an element of seriousness evident – and intended – in his superficially jocular remark that if the new engine failed to deliver the power which had been promised he would be packing a Ford DFV in his suitcase before making the journey to South Africa for the first world championship race in March.

The third absentee from the GP scene in 1969 was the Matra team. The French company, whose management were nothing if not realists, had taken a long, hard look at their engine and had decided after consultation with their engineers that the poor power output achieved so far could not be improved significantly without a major redesign. The limitations imposed by the valve layout and combustion chamber shape constituted the major problem, but there were also other areas which would benefit from some redesigning, and in order that this work could be carried out with the minimum of interruption it was decided to withdraw the works GP team for one season, with the intention of reappearing with the revised engine in a completely new chassis in 1970. This was tough luck on Pescarolo, who had been hoping for his first full season of Formula 1 racing alongside Beltoise, while it meant that Beltoise himself had to cease being a team leader and accept the role of number-two driver to Jackie Stewart in Ken Tyrrell's team which was to be the sole Matra representative in grand prix racing for the 1969 season, with the full backing of the French company's technical personnel and resources.

Pedro Rodriguez had been left out in the cold by the BRM reshuffle, but was invited to drive Tim Parnell's P126, for which there was the promise (not to be fulfilled) of one of the new four-valve engines, while Piers Courage, who had driven this car in 1968, had joined forces with private entrant Frank Williams to run a Brabham-Ford similar to the BT26A works cars, but on Dunlop instead of Goodyear tyres. Vic Elford was also scheduled to join the ranks of Ford-powered drivers as his entrant, Colin Crabbe's Antique Automobiles, had decided to replace their Cooper-

Maserati with a McLaren M7B chassis with pannier-type fuel tanks which Bruce McLaren intended to sell after the South African Grand Prix.

By the beginning of 1969 well over 30 DFV engines had been built, including the initial batch of seven completed in 1967, and nearly as many again were scheduled for production during the next 12 months, the latest versions having cylinder head modifications enabling the rev limit to be raised from 9,500 to 10,000 r.p.m. The vast majority of the engines were still being ordered by various Formula 1 teams, although for the second year running Lotus were to tackle the Tasman Championship with 2½ litre DFW-engined cars, and were joined on this occasion by a Frank Williams Brabham for Piers Courage, who had performed so well the previous year in his Formula 2 McLaren-Ford. This time there were only seven rounds of the Tasman contest, four in New Zealand and three in Australia, and Courage managed to win one of them while Rindt scored in two more. But this was to be Ferrari's year as Chris Amon, who ran a two-car team for Derek Bell and himself, clinched the title with four outright wins and two third places.

The DFV's sports car programme, meanwhile, had been relatively short-lived. The P68 had shown a lot of potential on its first appearance at Brands Hatch, and some of this had been seen again in subsequent appearances at the Nurburgring, Spa-Francorchamps, Silverstone and Oulton Park. But the car had been let down by a number of minor teething problems, and there was a severe setback to the development programme when Chris Irwin was critically injured in a high-speed accident while practising at the Nurburgring, his car being destroyed after flipping over on one of the notorious jumps. The car's last appearance in 1968 was during the Tourist Trophy, but it was to reappear in 1969 with larger brakes and a number of other refinements including a full-width rear wing, together with a completely new P69 sports-prototype with open-top bodywork. A major function of this car was to act as a mobile test bed for a programme of aerodynamic studies, and for this purpose it was fitted with a pair of interconnected front and rear airfoils, both of which were incorporated in the overall body shape and were adjustable both mechanically and hydraulically. But for an engine shortage the P69 would have appeared alongside the P68 in the BOAC 500 endurance race at Brands Hatch, but plans to enter it in later races were shelved when it became obvious that the major teams participating in endurance racing were preparing to build 25 special 5 litre cars in order to gain homologation under the Group 4 (later Group 5) Sportscar rules, which effectively ruled out the concept of a 3 litre car as a candidate for outright victory.

It was, perhaps, unfortunate that the P68 and P69 were not developed more fully, because they clearly had great potential, but the fact that neither sports-prototype was to be seen in action after May 1969 (when Frank Gardner's P68 was withdrawn from the Martini Trophy race because of the almost flooded Silverstone track after having set the fastest practice time) does not mean that the project was without value. There was, in fact, a direct link between the P68 and the GT70, the two-seater coupe which Len Bailey designed and built in great secrecy during 1970 and which was unveiled early in 1971 as Ford of Europe's prototype for a future rally winner and ultimately for a prestige high-performance road car. At no time, however, was serious consideration given to the idea of powering the GT70 with a DFV engine, although the thought was put forward on one occasion when lighthearted conversation penetrated the Ford boardroom!

Keith Duckworth, perhaps, was not unduly disappointed to hear that the sports-prototype's racing programme had been concluded, because in 1969 he found himself with very little time to spare for the sort of development work which might have been necessary had someone decided that his DFV engines would be required to run for 24 hours. His preoccupation at this time was not with engine design and development so much as with car design, for he and Mike Costin had finally succumbed to the temptation to build a grand prix car.

Like the majority of existing racing car designers they believed that the future revolved around four-wheel drive, and realizing that nearly everyone would be working in the dark, for the fund of practical experience of this type of transmission was very limited, they decided that this was the best possible moment to enter the field of racing car design. However, they needed a talented chassis designer with a real understanding of aerodynamics, and they managed to persuade Robin Herd away from McLaren Racing, with whom he had been employed since leaving the Royal Aeronautical Establishment research team at Farnborough, Hampshire.

The Cosworth four-wheel-drive car, however, was destined never to race. At one stage the project was halted when it became obvious that everyone was turning to wings. It was then reinstated when the safety legislation was introduced during 1969 to limit the size and type of airfoil aids, but finally it was abandoned completely towards the end of the year when knowledge gained from testing the Cosworth car convinced Duckworth and Costin that four-wheel drive was not, after all, an advantage under most racing conditions, nor in their opinion was it ever likely to be. After discontinuing their test programme with the car they carried through a transmission development programme involving an improved type of differential, but having provided themselves with the answers they were seeking they terminated this activity as well.

But Cosworth were not the only people to prove during 1969 that over-estimation of the potential of four-wheel drive was an expensive and time consuming mistake. Lotus, McLaren and Matra all produced and raced four-wheel-drive Formula 1 cars during the season, the Matra employing Ferguson transmission and the other cars transmissions developed in close association with Hewland Engineering, whose gearboxes were already featured on the vast majority of the world's racing cars. But not one of the 4WD cars was even to equal the performance of the two-wheel-driven cars in the same teams, and if anything their performance was least impressive in the wet, when the advantages of four driven wheels had been expected to be at their maximum. In the circumstances it was fortunate for Cosworth Engineering that the modifications which had been necessary to the DFV engine in order to mount it in a 4WD chassis were minimal, and amounted to little more than turning the engine back-to-front so that the clutch and gearbox were ahead of the engine and immediately behind the driver.

The relative failure of the four-wheel-drive cars in 1969 was due in considerable measure to the great strides which had been made in improving the adhesion of conventional grand prix cars. The tyre technicians had done a magnificent job in developing new tread compounds and methods of construction which not only provided more grip but were doing so at the considerably higher operating temperatures which had been imposed by the down-force created by wings. It was now so much easier to put 400 horse-power on to the road than it had been two years earlier when the first DFV engine was dropped into the original Lotus 49. Also, two unforeseen, or perhaps underestimated problems arose when the various four-wheel-drive cars were given their initial tests. The first was the considerable alteration in handling characteristics which occurred when the division of torque between the front and rear pairs of wheels was adjusted even slightly, and the second was the excessive understeer which was generated if the vast majority of torque was not fed through the rear wheels, thereby minimizing the theoretical advantages of the four-wheel-drive concept.

There was also the dual driving problems of adapting to a completely new technique and coping with the considerable increase in sheer physical effort needed to drive a four-wheel-drive car at anything like competitive speed, not to mention the virtual impossibility of finding sufficient time to develop either the new chassis and transmissions or the necessary driving techniques in the middle of a busy world championship race programme in which the main objective and preoccupation was the accumulation of points. In retrospect, therefore, it is not too difficult to understand why the four-wheel-drive movement in grand prix racing was so short-lived.

The first cars appeared in public for the first time at Zandvoort, although neither the Lotus 63 (based on the Type 64 Indianapolis car) nor the Matra MS84 (similar to Matra's new MS80 two-wheel-drive Formula 1 car, but with a spaceframe instead of a monocoque chassis) were to be used in the Dutch Grand Prix, Hill and Stewart having satisfied themselves during practice that they were achieving little with them beyond arm-ache.

John Miles was given the job of driving the Lotus 63 in its first race, the French Grand Prix, but his mechanical fuel pump packed up on the first lap, and it was not until the British Grand Prix that four and two-wheel-drive cars were to be seen racing against each other. The reluctance of the regular Lotus drivers to use the new cars had caused a storm of unrest in the team, culminating in Colin Chapman threatening to force the issue by selling off the Lotus 49s. But Rindt and Hill were still in their regular cars for the race, and the Lotus 63s were given to Miles and Bonnier. Beltoise found himself the somewhat reluctant driver of the Matra MS84 after Stewart had crashed his own MS80 during practice and taken over his team-mate's car, while the new McLaren M9A was in the hands of Derek Bell.

As a demonstration of the potential of four-wheel drive the Silverstone race was a failure, for only Beltoise and Miles finished the race, in the last two places, Bell retiring the McLaren with suspension damage after a puncture had sent him off-course a few minutes after the start, and Bonnier retiring immediately afterwards with a sick engine as a result of lubrication problems.

Mario Andretti's Lotus 63 was the only four-wheel-drive car in the German Grand Prix, and after two troublesome days had left him with the minimum of practice he started the race still unfamiliar with his car and crashed it on the very first lap, the accident also taking Vic Elford off the track with Colin Crabbe's McLaren M7B, the car being destroyed and Elford suffering a badly broken arm.

The somewhat startling-looking Cosworth car had been tried at Silverstone with the idea of running it in the British Grand Prix, but it had been withdrawn a few days before the race when it was obviously in need of more development, and in view of the poor showing by the other cars it seemed unlikely that Cosworth's plans to race their car in selected events would be sustained, although Mike Costin later gave a public demonstration of the sharply-styled car at a Ford Sport Speed Day at the Mallory Park circuit.

The McLaren project had been quietly shelved, meanwhile, so that only Lotus and Matra were continuing to carry four-wheel-drive cars in their transporter. John Miles was the lone representative on the starting grid with his Lotus 63 at Monza, but once again the car's suspect lubrication system brought an early retirement. He seemed to be having better luck in the Canadian Grand Prix at Mosport two weeks later, but a seized gearbox stopped him just before half-distance. However, Johnny Servoz-Gavin, who had been given the Ken Tyrrell Matra MS84, wrote a few lines of history by being the first driver to score a world championship point with a four-wheel-drive car, although sixth place was somewhat flattering for a car finishing six laps behind the race-winner.

Servoz-Gavin also finished in the United States Grand Prix, this time in seventh and last place and 16 laps behind after a long pit stop, but Andretti, back in the Lotus 63 again, started near the back of the grid after various practice problems, tangled with another back-marker immediately after the start, and retired soon afterwards with suspension damage. Finally, and it was to be the last occasion on which any of the 1969 four-wheel-drive cars were to be seen in action, Servoz-Gavin drove the Matra MS84 in the Mexican Grand Prix and put up the best performance of the season in finishing only two laps behind the winner in eighth place. It seemed a considerable achievement at the time, but it was modest enough compared with the aspirations a year earlier for four-wheel-drive cars as they were taking shape behind closely guarded doors in Hethel, Colnbrook, Northampton and Velizy.

But at least Matra could be content that not all of their work that year had been in vain. Their main effort had gone into the design and manufacture of a chassis to take the Ford engine, as a successor to the MS10 which Jackie Stewart had driven so effectively in 1968. Unlike the earlier car, the MS80 followed the familiar practice of bolting the engine directly to the rear of the cockpit bulkhead without the additional support of an underframe, but at the same time the Matra designers had taken advantage of the emergence of wings to create a body shape which was evolved around them. Hitherto, of course, wings had been mounted above or attached to bodies, the original aerodynamic characteristics of which had been achieved without resort to airfoils, and hence there had been an element of hit-and-miss about their subsequent size, shape and location. On the new Matra, the fins on the sides of the nose cone were tailored not only to the height, angle and size of the rear wing, but also to fixed fairings over the front suspension which blended into the sides of the body forward of the cockpit. Also, every effort had been made to concentrate the maximum amount of weight towards the centre of the car, and this had resulted in a relatively bulbous-looking body, albeit a most efficient one.

However, Stewart and Tyrrell, whose determination to avenge their narrow defeat in the 1968 world championship knew no bounds, decided to play safe and use their well-developed MS10 for the opening round in South Africa as it had proved to be slightly the faster during practice. Like so many of their decisions that year, it proved to be a wise one, because Stewart led the race from start to finish, to win comfortably from Graham Hill, Denny Hulme, Jo Siffert, Bruce McLaren and Jean-Pierre Beltoise. This meant that for the first time all six places had been taken by drivers and cars using the Ford DFV engine, an achievement which was to be repeated no less than five more times during 1969.

Jack Brabham, whose first appearance with the Ford engine had brought him pole position on the starting grid, tucked in behind Stewart on the first lap, but he was soon in trouble when one of his rear wing supports started to collapse. He stopped to have both the rear and front wings removed, but without them his car was handling too precariously and he was forced to give up, as was his new team-mate Jacky Ickx, who also stopped to have his wings removed and then found that he could not restart the car. Jochen Rindt was also out of luck in his first drive with the Lotus team, gradually slipping down the field after taking over second place from Brabham, and giving up before half-distance with a fuel pump failure.

With the South African Grand Prix moved from its traditional January date to March, there was a considerable rush after the race to return the cars to Europe, for although the next championship race, in Spain, was still two months away, the intervening period included the dates of the two British pre-season non-championship races at Brands Hatch and Silverstone.

In the first of these, the *Daily Mail* Race of Champions, Jackie Stewart drove his new Matra MS80 for the first time in a race, and gave a demonstration of the car's calibre by leading from start to finish, and being completely in command all the way. Jochen Rindt and Graham Hill provided the main opposition in their Lotus 49Bs, but Rindt dropped out with low oil pressure after setting a new lap record, and Hill was seven seconds behind at the finish, though a long way ahead of third man Denny Hulme.

For once the Brabham team did badly at Brands Hatch, Brabham himself retiring with ignition trouble and a fuel leak, and Ickx abandoning with a sticking throttle, but they turned the tables at Silverstone during the *Daily Express* International Trophy race, where Brabham held off a determined race-long attack by Rindt on the wet track to win by just over two seconds, coasting across the line with a dead engine following another fuel leak! Jackie

Stewart reverted to his earlier Matra MS10 for this race as the MS80 was still an unknown quantity in the wet (and was to remain so for a very long time as 1969 was to prove an unusually dry year), and he finished well back in third place, just over two seconds ahead of Ickx in the second Brabham.

The Brabham team's troubles at Kyalami had highlighted once again the potential hazards of wings, which by now were not only sprouting from both ends of cars but were flying higher than ever. But reservations regarding their desirability after the South African Grand Prix were as nothing compared with the outcry which followed the Spanish Grand Prix in May, during which both Graham Hill and Jochen Rindt suffered serious accidents as a result of wing failures, and Jacky Ickx was forced to stop when his own wing started to collapse later in the race.

For the first time the Grand Prix was held on the road circuit in Montjuich Park, on the outskirts of Barcelona, and fortunately the entire circuit had been lined with safety barriers before the race, following representations by the Grand Prix Drivers' Association, whose members had foreseen the potential danger to spectators lining the circuit in the event of an accident.

Lotus had arrived at the circuit with huge wings over their cars, and these had grown even larger before the start of the race, completely dwarfing Rindt and Hill as they lined up on the front row with Amon's Ferrari between them. Hill's accident occurred on the ninth lap, shortly after he had moved ahead of Siffert's Lotus into third place. Suddenly his Lotus went out of control over the hump a few hundred yards beyond the start line, at which point the cars are travelling in a slight left curve at well over 120 m.p.h. Without warning Hill's car slewed sideways, hit the barrier on the left, then rocketed across the track to hit the guard rail on the opposite side of the track, the metal screens in each case preventing the car from ploughing on into the crowd. The Lotus was wrecked, but Hill, though stunned, was unhurt. He climbed out, helped marshals to drag the wreckage closer to the barrier and picked up pieces which were still lying on the track. Then he started to examine the car.

At first he suspected a suspension failure, but he could find nothing that could not be explained by either one or the other of the two impacts. The only possible explanation, he decided, was that the wing had collapsed before the accident, and not because of it. He decided to rush back to the pits to warn Colin Chapman of the possibility of something similar happening to Rindt's car, which had been in the lead from the start. But scarcely had he left the scene of his own accident when Hill saw Rindt's car go out of control at the same spot, and this time there was no doubt about it – the wing had collapsed. Jochen's accident was much more serious, because his out-of-control car had catapulted off Hill's wrecked Lotus, and after mercifully having been guided in the general direction of the track by the barrier, had rolled over several times and finished upside-down, minus three of its wheels, at the side of the track. Hill rushed back to assist marshals in righting the car and lifting Rindt out of the wreckage, and he was immensely relieved to find him not too badly injured, though losing a lot of blood through a broken nose. In the circumstances Jochen had had a miraculous escape, and he was able to leave hospital within a few days, though he was to suffer nausea and occasional fits of giddiness for several weeks afterwards.

Rindt's accident had put Amon into a secure first place with the Ferrari, and he was driving smoothly to a comfortable victory when his engine failed at the start of his 57th lap, just as it had rather earlier in the race at South Africa. Jackie Stewart had been holding a firm second place behind the Ferrari, so suddenly he found himself with an overwhelming lead as several other drivers had run into trouble behind him; at the finish he had two clear laps over Bruce McLaren in second place, and three over Beltoise and Hulme. It was the first of many victories for the Matra MS80, but even Stewart had to admit that on this occasion he had been very fortunate to pick up his nine world championship points. At Barcelona the only car without a Ford engine to finish in the first six was John Surtees' BRM, six laps behind in fifth place following two pit stops to cure a persistent misfire which eventually was traced to a faulty spark box.

After all the trouble in Spain it was perhaps surprising that so many teams arrived at Monte Carlo ten days later with their cars still sprouting wings. But immediately after the first official practice session there was an emergency meeting of the CSI at which it was decided to ban wings from the Monaco Grand Prix pending the framing of new legislation covering all aerodynamic aids some time during the next few weeks. Because of the ban it was also announced that the times recorded during the first practice period would be disregarded for grid positions.

Both decisions caused something of an uproar amongst the teams, Ken Tyrrell being particularly vociferous in his condemnation of them, claiming that as his cars had been designed around their wings it would be potentially unsafe to race without them. It was a novel, if not entirely convincing argument, but it certainly carried no weight with the CSI representatives, who were on hand to see that the Matras, like all the other cars, were wingless for the next and all subsequent practice sessions.

Despite the lack of wings, Stewart was comfortably the fastest driver by the end of the third and final session and as at Barcelona Amon was the second quickest in the lone Ferrari, proving that the latest version of the V12 engine with its exhausts emerging from the sides instead of from the centre of the Vee, was at least delivering useful power even if so far it lacked stamina.

But poor Amon was destined for yet another retirement, this time on his 17th lap as a result of a differential failure when he was giving Stewart a hard chase from second place. But Stewart's race lasted only a little longer before he, too, had to give up with transmission failure, in his case involving a drive shaft universal joint. From then on it was Graham Hill all the way, to earn his fifth walk up the steps of the Royal box to collect a handshake and the winner's trophy from Prince Rainier and Princess Grace.

It was in this race that Piers Courage added greatly to his stature as a rapidly maturing grand prix driver when he drove Frank Williams' Brabham-Ford BT26A into a solid second place, just over 17 seconds behind Hill's Lotus 49B, and a similar distance ahead of Siffert's Lotus, having spent much of the race dicing closely with Ickx's works Brabham until it broke a rear hub carrier. Richard Attwood, who had done so well for BRM in the Monaco race in 1968, stood in this time for the injured Rindt, and gave Lotus fourth place ahead of McLaren and Hulme, Denny's lowly placing in a healthy car being explained by a curious throat infection which had left him extremely short of breath.

Following the Monaco wing decision it was interesting to observe the variety of aerodynamic solutions the following month when the grand prix teams met again in the Zandvoort paddock. Suspension-mounted wings were definitely out, by common consent, but not 'airfoils forming part of the bodywork', and there were a number of ingenious devices, ranging from the deep wing-shaped engine covers of the McLarens and Matras to the pure airfoils of the Lotus and Brabham cars mounted on open-ended boxes behind the engine. It was not until after this race that the CSI finally formulated new regulations, in which all airfoils in future were to be fixed and be considered as part of the car's bodywork. This meant that they were restricted to a maximum width of 150 cm. ahead of the front wheels, or 10 cm. if they were higher than the wheel rims, and 100 cm. behind the front wheels. Also, no part of them could be more than 80 cm. above the lowest point of the fully sprung structure of the car.

Rindt, meanwhile, had made a remarkable recovery from his Barcelona injuries, and demonstrated his growing fitness by making the fastest practice time for the Dutch Grand Prix less than seven weeks after his accident. This time Amon could manage only the fourth fastest time, behind Stewart and Hill, to claim a second-row starting position alongside Ickx.

It seemed that Jochen was determined at all costs to demonstrate that his practice time was no fluke, and at the end of the first lap he outbraked Hill at the end of the main straight and with the aid of a spot of grass squeezed past into the lead. Four laps later Stewart also went past Hill, but Rindt by this time was holding a clear lead, which he extended lap by lap until, when it had grown to over 10 seconds, he suddenly slowed down with a broken drive shaft joint.

Hill, meanwhile, had dropped back slightly, and Siffert, who had been barrelling his way through the field after a slow start, took up the chase. But there was nothing he could do about Stewart's Matra, which went on to win by nearly 25 seconds. This time Amon was in luck because at least the Ferrari finished the race, the New Zealander crossing the line six seconds after Siffert and nearly seven seconds ahead of Hulme, who just pipped Ickx for fourth place. Brabham was the only other driver to complete the full 90 laps.

1969 was to prove a very happy year for the Matra company, despite the absence of their own all-French Formula 1 team, but there was no occasion to equal for them the closing seconds of the

French Grand Prix, at Clermont-Ferrand, where Jackie Stewart won his fourth world championship race of the season in five starts, and Jean-Pierre Beltoise fought a wheel-to-wheel battle with Jacky Ickx's Brabham to take second place in the other Tyrrell Matra. For a French car, albeit with a British engine and with the help of a British-organized team led by a British driver, to take the first two places in France's own grand prix was just too much for the partisan crowd, who went almost berserk with delight while the local radio commentators were red-faced with passion and high blood pressure as they poured out their glad tidings to their listeners throughout the French-speaking world.

For Stewart it had been another start-to-finish demonstration, with only Hulme, who had joined him on the front row of the grid, able to offer any sort of challenge during the opening laps before dropping back with a loose front anti-roll bar which later he had to have secured at the pits. Ickx had taken over second place when Hulme made his pit stop but Beltoise, who never drives better than when in front of his home crowd, launched a sustained challenge and had closed to within a second of Jacky as they went into their 38th and final lap. The Frenchman was looking for an opening on every corner, and he found it during the second half of the lap when Ickx made a momentary error and ran wide. Beltoise was through in a flash, and although Ickx fought back magnificently, taking the swerves before the final corner at an alarming rate to close right up on to the tail of the Matra, Beltoise had the corner and was able to hold off the Brabham to claim second place by a bare fifth of a second.

Bruce McLaren, Vic Elford and Graham Hill all finished a lap behind, but once again Rindt and Amon were not to finish their race. Several drivers, including Stewart, had been upset by the continual undulations and swerves of the magnificent five-miles Clermont circuit, but it had affected Rindt, still suffering some of the after-effects of his Barcelona accident, most of all and he had been forced to quit shortly after passing the half-way point in fourth place while Amon, who inherited his place, dropped out seven laps from the end with yet another engine failure.

By now it was fairly clear that Jackie Stewart was driving serenely towards his first world championship, for the combination of his driving ability, the meticulous preparation of Ken Tyrrell's devoted mechanics, and a competitive car and engine was proving devastatingly effective. Stewart's task had obviously been made easier by Jochen Rindt's accident in Spain, for although he had made a remarkable comeback he was still not 100 per cent fit.

Yet there was little evidence of that fact at Silverstone, where the two drivers, who were near-neighbours and close friends at their adopted home in Switzerland but razor-keen rivals on the track, put on a brilliant display of on-the-limit driving in the British Grand Prix. Stewart had been very fortunate even to start the race, because during the final practice period, when battling against Rindt for the fastest-lap prize, he had hit a piece of concrete which had been dislodged from the retaining wall lining the corner before the pits by another car and was lying in his path. The Matra spun twice, and Stewart slid helplessly into the tall bank in front of the main grandstand, emerging slightly shaken from the badly creased car. He had immediately climbed into Beltoise's car, in order to qualify it, while Jean-Pierre set about accustoming himself to the peculiarities of the Matra MS84, and in the remaining minutes of practice he managed to set a lap time second only to Rindt's best with the Lotus. It was a formidable effort.

Right from the start the race developed into a private contest between Rindt and Stewart, with Hulme doing his best to hang on in third place until slowed and eventually stopped by engine trouble. The Lotus led for the first six laps, and then Stewart squeezed past in the Matra, but only for another nine laps, before Rindt forced the Lotus back in front again. They duelled away for lap after lap, rarely more than two seconds apart, until suddenly after 62 of the 85 laps Rindt shot into his pit. One of the vertical end blades of his wing had been rubbing against his left rear tyre through right-hand corners and had become buckled. Rindt was understandably worried that the jagged edge of the wing would damage the tyre and he rushed in to have the offending part cut away.

He shot off again, still in second place, but the gap between the two cars was now more than 20 seconds so the battle effectively was over, and Stewart's victory became virtually assured when a very annoyed Rindt had to stop again a few laps from the end of the race to take on more fuel. Hill and Siffert also ran short during the closing laps and had to make unscheduled stops for a fill-up (the Lotuses had been running without their supplementary fuel tanks), and the long race also nearly cost Jacky Ickx his second place, a lap behind the Matra. Despite dropping his engine revs in reply to pit signals, Ickx heard his engine splutter half-way through his final lap, and he just managed to coast his Brabham across the finish line to secure his place six seconds ahead to Bruce McLaren, a drama which Jack Brabham had watched in agonized silence from his pit, one ankle heavily bandaged following a testing accident at the circuit several weeks earlier. Rindt went on to take fourth place after his fill-up ahead of Piers Courage and Vic Elford, which was less than a just reward for his great drive.

The British Grand Prix had brought the first part of the 1969 world championship series to a close, and Jackie Stewart was one more long step closer to his first title. He now had an impressive 45 points from six races, having failed to collect a maximum score only at Monaco. The consistent Bruce McLaren was running in second place, but with only 17 points, one more than Graham Hill. The battle was not yet won, but in military terms only mopping-up remained to be done.

One more victory, in the German Grand Prix, would have clinched the title for Stewart, even though there were still four more races to be run, but it was not to be. Jacky Ickx, whose knowledge of the Nurburgring's 170-odd corners was already prodigious, gave notice that he had picked up the scent of victory by setting the fastest practice time, just three-tenths of a second quicker than Stewart's best, and almost eight seconds quicker than Rindt's third-fastest time. There were only 13 Formula 1 cars entered for the Grand Prix, and so the numbers on this occasion were made up by eight Formula 2 cars, whose drivers were to compete for separate awards.

It was the second time running that Ickx had set the fastest practice time for the German Grand Prix and it was also to be the second time running that his supreme effort was to be wasted by a slow start. Questioned about this after the race, he denied that he had been slow away from the line, but conceded that maybe everyone else around him had been a bit faster! Stewart was driving the car which he had crashed before the British Grand Prix, and he rocketed straight into the lead to complete the first lap comfortably ahead of Siffert and Rindt, with Ickx now in fourth place having courteously been signalled to pass by two or three other drivers who knew that he was the only person likely to stop yet another Stewart win!

At the end of the second lap Ickx was ahead of Rindt and in third place, and with three laps on the score board he was up into second position ahead of Siffert; now he set his sights on Stewart's Matra, which he could see ahead of him at several points on the circuit. Within two laps the cars were nose-to-tail, but it was not until the start of the seventh lap that the Belgian was able to nose his Brabham in front and keep it there. But from that moment on he dominated the race completely, his lead being extended comfortably as Stewart ran into intermittent gear-selection trouble. By the end he was almost a minute ahead, and only McLaren and Hill were still mobile in Formula 1 cars at the finish, although Siffert and Beltoise were classified after both had stopped with damaged front suspension late in the race. August was to prove a fine month for Ickx, for two weeks later he also won the non-championship International Gold Cup race at Oulton Park, ahead of Rindt, who had his one and only race in the Lotus 63.

Stewart had been desperately keen to secure his world championship with an outright race win, and even though now he needed only a few points to put the title beyond mathematical doubt he still wanted to do so with one more outright win. In many ways, Monza is the most difficult place to prepare for this because no matter what you do as a driver slipstreaming counts for so much, and more often than not the Italian Grand Prix is won and lost in the rush up to the finishing line of the final lap. However, Stewart and Tyrrell went into a huddle, and decided to use gear ratios specifically tailored to that final sprint from the *Curva Parabolica* to the chequered flag.

Jackie had ample opportunity to prove that their gearing calculations were correct during the race, and it was significant that although the leading bunch of cars, which had been reduced to four by the finish, were chopping-and-changing places all the way around the circuit, Stewart's Matra actually crossed the start-finish line in the lead on 60 of the 68 laps.

At the final corner the battle was between Stewart, Rindt, Beltoise and McLaren, and this is how they crossed the line, the gaps between them as they took the flag being just eight, five and two hundredths of a second respectively. Stewart had driven an exhausting race to achieve his great ambition, but the seeds of his success had been sown in the Matra garage the night before the race when those gear ratios were checked and checked again. Piers Courage should have been a fifth driver battling it out to the chequered flag, but a few laps from the end his Brabham developed an ignition fault, and he lost his tow, although he was able to complete the full race distance, two laps ahead of the next man, Pedro Rodriguez, who had been the sole representative of Ferrari that day.

Chris Amon, whose last race for the Italian team had been in the British Grand Prix, where he had retired with gear-selection problems, had been pinning his hopes on the brand new Ferrari 312B and its new flat-12 engine which had shown some most impressive output figures on the test bed at Maranello. The week before the Italian Grand Prix he had been at Modena carrying out tests on the car prior to unveiling it at Monza, but although it seemed to be a vast improvement on the previous model, and the engine much more lively, the test session was cut short by a whole series of mechanical disasters, and a very despondent Amon watched practice for the Grand Prix from the pit road, content to let Rodriguez, who had been his team-mate at Silverstone, do the driving in the old and hitherto unreliable car. The development problems with the new Ferrari were to continue for some months, and in fact Amon was destined not to drive for the team again. Within a few weeks of the Italian race he had agreed to join the newly formed March team, although had he been able to foresee the remarkable upswing in the Ferrari team's fortunes during the second half of the 1970 season he might well have been less keen to leave Italy.

Although Stewart had won the championship, there was no let-up in the Tyrrell team's effort as they prepared for the three remaining Transatlantic grands prix. The Canadian race this time was back at Mosport, and practice times amongst the leading cars and drivers were remarkably close until Ickx suddenly lapped a full half-second quicker than anyone else, a bare fifth of a second covering the times of the next five cars.

Ickx had shown a certain amount of temperament in the Brabham pit at Monza when things had been going badly for him during practice, and just before the race in Canada he had announced his intention to leave the team at the end of the year in order to return to Ferrari, news which scarcely helped to cement relations with his current masters. But there was no doubt that he was in close harmony with his car, and the personal recriminations were all forgotten after the race when the smiling young Belgian returned to the paddock having won the Grand Prix. But it was also a humble Ickx, who had taken time off on his way back to the paddock to make a public apology to Jackie Stewart over the public address system for a driving error which had resulted in him nudging Stewart's car off the road and out of the race.

It was an unfortunate incident in an otherwise brilliant drive, with Ickx and Stewart locked in close combat from the eighth lap until the moment when Ickx made his faulty overtaking manoeuvre on lap 33. With Stewart out of the race there was no one to worry Jacky, for Rindt, who had led initially, was more than half a minute behind in second place and gradually losing ground, and in fact was caught and passed by Jack Brabham later in the race. This was only the second occasion on which the Canadian Grand Prix had been held at Mosport, and the second time that the Brabham team had finished first and second. Rindt was the only other driver to complete the 90 laps, Beltoise being one lap behind in fourth place and McLaren finishing fifth, a further two laps in arrears.

Jochen Rindt had waited a long, long time to win his first world championship grand prix, but he could scarcely have chosen a better race than the United States Grand Prix which in 1969 carried a first prize of 50,000 dollars. The manner of Rindt's victory was reminiscent of Stewart at his best, and once again these two drivers were to decide the outcome of a grand prix on their own. However, Lotus joy that day was marred by Graham Hill's serious accident late in the race following a punctured rear tyre, as a result of which he badly damaged both legs.

Rindt had set a practice time fractionally quicker than Hulme's best to secure pole position, and Stewart was right behind him and alongside Hill on the second row. But as Hulme was in gear-selection trouble even before the race started (his mechanics worked furiously on his car when the start was delayed as officials tried to clear the track of wandering spectators) he could not be considered a threat, while his poor team-mate Bruce McLaren was already viewing the race preparations from the bridge overlooking the start, having blown-up his car on the warming-up lap.

When the flamboyant Tex Hopkins did his usual purple-suited leap into the air with his starter's flag, Rindt swept unchallenged into the lead as Stewart tucked in behind him and Hill slotted the other Lotus into third place. Even before the end of the first lap the leading Lotus and the closely following Matra were pulling out a sizeable gap on the rest of the field, and they ran together nose-to-tail until the twelfth lap when Stewart slipped past. But he was not to lead for long, for Rindt was soon counter-attacking, and he went back into first place on lap 21. Soon Jochen had opened up a useful lead, and the reason became obvious very shortly afterwards when wisps of oil smoke emerged from the back of Stewart's car. The smoke gradually built up into a cloud, and after 35 laps Stewart brought the Matra in to retire, the back of the car bathed in oil.

Second place was now being held by Piers Courage, who was driving his Brabham magnificently to hold off a determined challenge by Ickx and Brabham in their works cars, a performance which was to earn him the award of Man of the Race as well as six championship points. This time Ickx was out of luck, retiring with engine trouble with just over half an hour of racing to go, and Brabham, who had tried repeatedly to get past Courage but without avail, had to take on more fuel a few laps from the end when one of his tanks failed to scavenge properly.

During his stop Brabham was passed by Surtees, whose third place was to give BRM their best result of a depressingly lean year. A whole succession of mechanical troubles, the inability of anyone to overcome them, and deteriorating personal relationships had brought the team's morale to a new low ebb. Chief engineer Tony Rudd had been dismissed at short notice in the middle of the season but this seemingly had had little effect on the team's performance except that they had missed the French Grand Prix, and Surtees had been in so much trouble with his car before the German Grand Prix that he withdrew it from the race. His third place at Watkins Glen, therefore, was a badly needed morale-raiser, although it did little to cheer Jackie Oliver whose engine once again had let him down, and who understandably was upset that he had yet to pick up a single championship point in his first year with the team. As it transpired, he was to do precisely that, in the one remaining race of the 1969 season.

After two lean grands prix, Stewart was impatient to add just one more win to his collection of six victories, but other drivers were equally hungry for the sight of the chequered flag, and at Mexico City the new world champion had to be content with fourth place, just one position in front of his team-mate Beltoise. During practice, Brabham and Ickx had shown the competitiveness of their cars in the rarefied atmosphere of Mexico City by setting the fastest times and filling the front row of the grid, while Denny Hulme had been less than a tenth of a second slower than Stewart to line up alongside him on the second row. A tough battle between these four was indicated, and indeed was to be the main feature of the race.

Stewart made an excellent start and led for the first five laps, but Ickx, Hulme and Brabham passed him one at a time so that he was relegated to fourth place by the end of the ninth lap. By this time Hulme had also passed Brabham and was challenging Ickx very hard for the lead, and he went through into first place on the following lap and was never headed again. It was his only Formula 1 victory of the year, but Hulme was in tremendous form that day, and he held off a race-long challenge by Ickx to lead him across the finishing line by two-and-a-half seconds. Brabham was just over half a minute further behind at the end, and Stewart another eight seconds in arrears, having been unable to match the pace of the leading cars with their latest-pattern Goodyear tyres.

Stewart's final points tally for the year was an impressive 63, but only 18 of them had come from the last five races, whereas Jacky Ickx had been able to accumulate 24 during the same period, giving him an end-of-season total of 37, which put him securely into second place in the table. Bruce McLaren had finished the season in third place with 26 points and Jochen Rindt was fourth with 22, a suspension breakage in the Mexican Grand Prix having prevented the United States Grand Prix winner from adding to his score.

The season had assumed a somewhat lop-sided look during the early part of the year as Stewart had notched up win after win in a most impressive display of driving and team efficiency, but since securing the championship Ken Tyrrell's team had felt the frustration of failure on more than one occasion, as a result of which the season as a whole had ended with a more balanced scoresheet.

The first four places in the championship table were secured with different makes of car – Matra, Brabham, McLaren and Lotus, but the impressive common denominator was the Ford DFV engine which had powered them all. Indeed, the Ford engine had been fitted to the cars used by all of the top ten drivers in the championship, as well as four of the remaining eight point-scorers, and of the grand total of 275 world championship points amassed by drivers during 1969, all but 14 of them had been won with Ford power. This was a measure of the contribution which the DFV engine had made to grand prix racing during its second season, and of the extent to which it had come to dominate it.

10

1970: Ambition fulfilled

EVERY NEW SEASON brings renewed interest to the grand prix scene because invariably it is accompanied by the announcement of new cars or improvements to the previous year's models and by news of new driver-team associations. But rarely had there been as many changes as took place before the start of the 1970 grand prix season.

There were several reasons for this, amongst which one of the most significant – and surprising – was the establishment of a completely new racing car manufacturer, March Engineering, who not only intended to offer grand prix cars for sale, but would also be operating their own works team. The company had been formed by Max Mosley (in charge of finance and publicity), Alan Rees (in charge of manufacture), and Robin Herd (who left Cosworth Engineering to become chief designer), and the programme which they announced during their first press conference in September 1969 sounded more than a trifle optimistic. But the March company was to achieve far more during the first year than their ambitious directors had thought possible, and as a result quickly became a most important member of the motor racing industry.

Chris Amon, having decided that it was time to move on from Ferrari, accepted March's invitation to lead the works team and he was joined by Jo Siffert who, despite a very cordial racing association with Rob Walker, was very anxious to secure a place in a works team. A third works-assisted car was to be run for the young Swedish star Ronnie Peterson, although it would be entered under Colin Crabbe's Antique Automobiles banner, and then as part of March Engineering's sponsorship tie-up with the STP Corporation of America it was revealed that STP's Indianapolis-winner, Mario Andretti, would also be racing a March car during the 1970 season when his American racing commitments did not clash with the Formula 1 programme.

But the icing on the March Engineering cake was provided by the news that Ken Tyrrell and Jackie Stewart had also decided to run a team of March cars. It was rare enough for a new company to burst straight into grand prix car manufacture without first having flexed their muscles for a year or two producing and racing less sophisticated racing cars, but it was unique that one should also manage to get the current world champion and his entrant to sign on the dotted line even before the first Formula 1 car had been constructed, far less turned a wheel.

It was a measure of the confidence being shown in the new organization, and perhaps a vindication of the company's decision to produce a completely conventional car for their first season of grand prix racing in the hope that they would achieve maximum reliability even though they might not achieve 100 per cent competitiveness.

Tyrrell and Stewart had been forced to go shopping for a replacement for their Matras because they were determined to continue racing with the Ford DFV engine which had served them so well in 1968 and 1969, whereas Matra had announced that they would no longer allow any of their Formula 1 cars to race with anything other than their own V12 engine. Before making his decision, Stewart carried out some tests with the much-modified Matra engine installed in one of his works cars, but although it was obviously a considerable improvement on the original V12 its performance caused him to be more convinced than ever that he was right to stick to Ford power.

The parting of the ways by Tyrrell and Matra meant that Jean-Pierre Beltoise was able once again to accept the job of number-one driver of the re-vamped French team, who were to run completely new cars called the MS120 with very distinctive, if rather ugly and angular bodies. Henri Pescarolo was retained as the driver of the second car and Johnny Servoz-Gavin was invited by Ken Tyrrell once again to become Jackie Stewart's back-up man, this time driving one of the new March-Ford 701s.

Stewart's closest rival in the 1969 world championship, Jacky Ickx, was now the new team leader at Ferrari, and clearly he had more faith in the 312B chassis and its flat-12 engine than his predecessor. Although Ferrari intended to run only one car at the beginning of the year, the plan was to operate a two-car team for the majority of the races, and both Clay Regazzoni, a rapid Italian-Swiss who had emerged through the ranks of Formula 3 and Formula 2, and Ignazio Giunti, a young Italian who had already established himself as a fine sports car driver, were signed-on by Ferrari with the intention of using them alternately in the second grand prix car.

Ickx's move to Ferrari had left a vacancy in the Brabham team, and this was filled by the German driver Rolf Stommelen, who was making his entry into Formula 1 racing with the aid of backing by the magazine *Auto Motor und Sport.* The monocoque-chassis car which Ron Tauranac had placed under a dust sheet early in 1969 when he found that the revised BT26 went so well with the Ford DFV engine was retrieved to become the prototype of the BT33 1970 Brabham Formula 1 car.

The McLaren team of Bruce McLaren and Denny Hulme were starting the new year with an improved chassis, the M14A, as a replacement for the previous year's M7A and M7C, the latter car with its more enveloping monocoque chassis having been sold to John Surtees who intended to use it pending the completion of his own Ford-powered Formula 1 car. But the works McLaren team also decided to run a third car experimentally with an Alfa Romeo engine developed from the company's V8 sports car power unit, and Andrea de Adamich, now fully recovered from the injuries he had sustained when he crashed his Ferrari at Brands Hatch in 1968, was named as the driver of this Anglo-Italian project.

Pedro Rodriguez may have had a lean 1969 season, but he had burst back into the limelight as the new team leader of BRM, with Jackie Oliver remaining with the team as his back-up driver. A major reorganization had taken place within the BRM team following their various troubles of the previous year, and a completely new chassis, the P153, had been designed by Tony Southgate, while Aubrey Woods (like Southgate a former AAR Eagle-Weslake man) had been entrusted with the task of developing more power and reliability from the BRM V12 engine. Much of the team's lost morale had been regained by the promise of a much more competitive car than they had known so far under the 3 litre formula, and there was a further boost to the team's spirits when it was announced that it was to be sponsored by the Yardley cosmetics concern and run in their gay colours instead of the rather drab green which had been used by BRM for so many years. A third formula 1 car was to be entered for the wealthy young Canadian, George Eaton, who wanted some grand prix experience and was prepared to pay a lot of money for it.

John Miles' sterling work with the four-wheel-drive Lotus in 1969 was rewarded by a regular place as number-two driver to Jochen Rindt in Gold Leaf Team Lotus, while Graham Hill, still recovering from his serious Watkins Glen injuries, accepted an invitation from Rob Walker to take over the Lotus cockpit vacated by Jo Siffert, the arrangement to have the backing of Brooke Bond Oxo sponsorship. For the early 1970 races, Lotus were to rely on their updated Type 49C cars, but the wedge-shaped Lotus 72 was known to be on the way, and was expected to be an extremely competitive car.

Meanwhile another newcomer to grand prix racing had appeared in the form of the de Tomaso 505, yet another car with a Ford DFV engine, which the Italian company had entrusted to Frank Williams to race on their behalf with Piers Courage in the cockpit. Like the March, the de Tomaso was a relatively conventional design, and as a result its initial building programme was completed very swiftly enabling the car to take part in the first of the season's Formula 1 races in South Africa.

A quick look at the entry list for the race at Kyalami provided vivid evidence of the extent to which the resources of Cosworth Engineering would be tested during the coming season. Of the 23 cars which lined up on the starting grid, no less than 17 of them were powered by the DFV engine, and there were probably at least an equal number of back-up engines lying in the respective teams' transporters in the paddock. To maintain all these and other engines, as well as manufacture a further batch of new units and if possible carry through more development work on the DFV, was a tremendous undertaking for what was still a relatively small organization.

Some of the engines must have been fairly tired even before the

South African race began because several of the teams had been spending many days at the circuit carrying out tests with their new cars and their respective tyre suppliers' latest products. It is mainly through this behind-the-scenes-work, well away from the crowds and whenever possible beyond the scrutiny of rival teams, that most of the progress in racing car performance is achieved, although no matter how long and how hard they have tested their car it is not until entrants and drivers come face-to-face with each other during official practice that they are able to evaluate the success or otherwise of their testing programmes. It is one thing to slice some time from the previous year's lap record, but it is only during the days preceding the first race that you know for sure whether you have been able to cut yourself a larger slice than any of the other teams.

Measured in these terms both the Tyrrell and the works March teams must have felt that their testing time at Kyalami had been well-spent, because Stewart and Amon were able to snatch the first two places on the starting grid with their March 701s with identical times, while Jack Brabham seemed about to demonstrate the competitiveness of his year-old 'new' car with its DFV engine by securing the third place on the front row. Ickx's Ferrari, fifth fastest in practice, was alongside Rindt's Lotus on the second row, and Beltoise managed to place the faster of the two new Matras – now known officially as Matra-Simcas in recognition of the company's new association with Chrysler – on the third row alongside the McLarens of Hulme and Surtees. The fastest BRM, however, with Oliver in the cockpit, was no higher than the fifth row.

In 1969 Jackie Stewart had led the South African Grand Prix from start to finish, and for a while it looked as though he might do the same thing again, especially after an incident on the very first corner had delayed several of the cars and drivers most likely to challenge him. Jochen Rindt had entered the corner very quickly with his Lotus, cannoned off Amon's March and then ridden over one of Brabham's wheels before sliding off course and dropping nearly to the back of the field.

Of the three cars involved, Amon's retired quite early with a split in the neck of its water filler (although as Andretti's car also dropped out with similar trouble it probably had nothing to do with the first-lap incident) and Rindt's finally succumbed with a rough-sounding engine after a hard fight back through the field from 16th place to fifth.

Only Brabham seemed unaffected by the leap-frog incident, and after assuring himself during the first three laps that his car was undamaged he started to apply the pressure, moving up from sixth place to second during the next three laps by passing McLaren, Oliver, Beltoise and Ickx. He then closed in on Stewart's March, and after several attempts to pass finally found a way through on lap 20, exactly quarter-distance, and remained in front all the way to the chequered flag. Stewart was finding it quite hard work trying to keep up with the new Brabham BT33, which clearly was handling a lot better than the March, and he had to give way to a determined attack by Hulme shortly before half-distance and resign himself to third place. Just over 17 seconds covered the first three cars at the end of the 204-miles race, and the only other driver to complete the full 80 laps was Beltoise in the Matra-Simca.

But the main opposition to the Ford-engined cars had been provided by Jacky Ickx until he was forced to retire his Ferrari with a damaged sump after glancing off Siffert's spinning March. Ickx had been running second to Stewart during the early laps, but the car had then faded slightly and was running in fifth place when it was forced out of the race.

Graham Hill was one of the stars at Kyalami, not because he finished sixth in Rob Walker's Lotus 49C behind John Miles' works car, but because he took part in the race at all. Hill had not long left hospital, was still very weak on his bandy legs, and had undergone great pain in order to so speed his recovery from his serious injuries as to be able to take part in the first race in the 1970 grand prix calendar.

He received an equally warm reception at Brands Hatch two weeks later where as usual the European part of the season opened with the *Daily Mail* Race of Champions, in which he finished in fifth place. This was a memorable race, for it provided the first victory for the March 701, although in very fortunate circumstances. Jack Brabham had shown throughout practice and most of the race that the form which the BT33 had shown at Kyalami was no fluke, and that it was probably the best-handling Formula 1 car of all at that time. But three laps from the finish, when holding a commanding lead, Brabham's spark box failed and he dropped to fourth place while making a pit stop for a replacement. Stewart went on to win the race comfortably from Rindt and Hulme, while Peter Gethin brought up the rear in his first drive in a works McLaren Formula 1 car, his reward for taking the Formula 5000 Championship the previous year in a works-backed McLaren M10A. The Race of Champions had also been notable for offering the first evidence of the growing competitiveness of the new BRM P153, with which Jackie Oliver had been one of the fastest in practice and had then led the race for eight laps before retiring with a transmission problem.

But the BRMs were to have more serious troubles the following month when the world championship battle was resumed at Jarama, outside Madrid. A series of stub-axle problems during practice was followed by Oliver leaving the track on the opening lap of the Spanish Grand Prix and colliding with Ickx's Ferrari as the BRM re-entered the track beyond the next corner. Both cars exploded into flames and burnt to a cinder, and although Oliver had a miraculous escape with nothing worse than singed eyebrows, Ickx was trapped in his car for a few seconds and received some nasty burns which kept him in considerable pain for the next few weeks whenever he climbed into the confined cockpit of a racing car.

Prior to the accident the meeting had been memorable mainly for the utterly chaotic organization, which had resulted in complete confusion throughout most of the practice periods as race officials repeatedly changed their minds as to whether specific sections of practice would or would not be timed or counted as qualification periods. Several drivers were not sure if they had qualified for the race or not, even as the cars were being lined up for the start, and there were even some brawls between understandably irate drivers and heavy-handed officials who tried to bar them from the grid.

The only driver who left the circuit with a smile that day was Stewart who, although he had managed only the third fastest time in practice, found that for the second year running the sun shone on him in Spain. Brabham and Hulme had taken the first two places on the grid, but Stewart emerged in front after the Oliver-Ickx accident, and remained there throughout the race. The track surface in the vicinity of the burning cars was made as slippery as an ice rink as fire-fighters went about their work in a most amateurish way, and Hulme bequeathed his second place to Brabham after nine laps, having slid helplessly off the track. But a few laps later Brabham also spun at the same point and dropped behind Beltoise into third place again. He recovered quickly, however, and went after the Frenchman in a very determined manner and was just about to pass him into second place when the Matra-Simca's engine seized. Jack then set about closing the gap behind Stewart, and within 20 laps had virtually cancelled out Jackie's lead of six seconds. The two cars ran almost nose-to-tail for many laps, and Brabham was obviously biding his time. But then, just when he started to attack really hard, he suddenly pulled up with a sick engine, and from then on Stewart was completely unchallenged as McLaren, who was now in second place, was already over a lap behind. With Mario Andretti finishing third and Johnny Servoz-Gavin fifth and last, it was a fine result for the new March company, while once again Graham Hill had managed to collect points, his steady drive being rewarded this time with fourth place.

Poor Chris Amon had been out of luck once again, this time his clutch failing as he left the start line, but the New Zealander was about to win his first Formula 1 race after a patient wait of eight years. It happened at Silverstone the following week, where the *Daily Express* International Trophy race was run in two parts, with Formula 5000 cars being allowed to compete alongside the grand prix cars. Amon had been comfortably the fastest during the mainly wet practice periods, and he went straight out in front at the beginning of the first part of the race, building a lead of just over 12 seconds by the end of it. Jack Brabham was his nearest challenger for 23 of the 26 laps until he pulled up with a damaged engine, after which Jackie Stewart took up the chase in a car which was being slowed by an excess of understeer. As Piers Courage was quite a distance further behind in third place with the de Tomaso-Ford, only Stewart seemed to have any real hope during the second part of the race of robbing Amon of the victory which he wanted so desperately.

Jackie set off on a wet and very slippery track at the beginning of the second half, determined to eat into those 12 seconds of lead, and was soon putting the March into some lurid opposite-lock slides.

But Amon kept a very cool head, and by driving in a much less spectacular and smoother fashion he was able after a few laps to match Stewart's times and then reduce the gap a little so that by the end of the race the two cars were only two seconds apart; Amon had won on aggregate by ten seconds, and his long wait for a Formula 1 win was over, although naturally he would have preferred it to have been accompanied by some world championship points!

Racing is full of 'ifs' and 'buts', perhaps too full, but it is probable that after the Monaco Grand Prix, the next race in the championship series, Amon reflected that had he not suffered a collapsed rear suspension on his March when lying in a comfortable second place behind Jack Brabham with three-quarters of the race behind him, he might well have shaken hands with Princess Grace that afternoon. For the Monaco Grand Prix of 1970 will long be remembered not as the race which Jochen Rindt won but as one which Jack Brabham lost – on the very last corner.

Brabham had begun the race in third place and had found a way past Amon on the 22nd lap. Six laps later he suddenly inherited first place from Stewart, who had led from the start from pole position but had now rushed into his pit with a misfiring engine to have the spark box and rotor arm replaced, losing two laps in the process.

From then on Amon appeared to be Brabham's only real threat, and the two cars were still only ten seconds apart when a bolt dropped out of the March's rear suspension. But Amon's ill-fortune spurred Jochen Rindt, hitherto in third place with the Lotus 49C, into mounting a tremendous challenge during the final quarter of the race, ahead of Siffert, Pescarolo and Hulme. With 20 laps to go Brabham held a lead of less than 14 seconds over Rindt, and if anything Siffert was closing up on Jochen, although he was soon to drop back again as his engine became starved of fuel. With 10 laps remaining the gap between the first two cars was still 10 seconds, and Brabham seemed to have the situation well in hand. Another five laps and the gap had not been reduced appreciably, but then the leader was baulked by some slow-moving cars, notably Siffert's March which was barely drawing any fuel at all by this time.

Rindt realized he now had a chance of victory, however slim, and he drove harder than ever to make up those last few seconds. As they went into their final lap there was still 1½ seconds between the two cars, which is a lot of time around the tight Monaco circuit. But still Rindt didn't give up, and although Brabham still had at least a second in hand as they emerged from the tunnel on to the Monte Carlo promenade he was obviously worried by Rindt's presence, otherwise it is doubtful whether he would have overshot his braking mark going into the final corner for the last time. He had had to take a tight approach line in order to pass a back-marker, and he failed to allow the extra stopping distance required on this more dusty section of the track. Inevitably, the Brabham locked up its wheels, and Jack slid helplessly straight on into the metal barrier lining the course as Rindt quickly altered his line and slipped around the back of the stationary Brabham.

Jochen threw up his hands in disbelief as he crossed the finishing line on his own, a state of mind which apparently he shared with the official holding the chequered flag who failed to wave it at him! Poor Brabham, aided by a marshal, managed to extricate his car, and drove on with the nose badly crumpled to cross the line in second place, 23 seconds later. He was a bitterly disappointed man, though his six points for second place had regained him the lead in the world championship table by 15 points to Stewart's 13.

There was a sad atmosphere in the paddock at Spa-Francorchamps the following month when the teams met again for the Belgian Grand Prix after a two-years' absence. The winner of the race in 1968 had been Bruce McLaren, but neither he nor his team were present this time for Bruce had lost his life a few days earlier while testing his new Can-Am sports car at Goodwood. He was, perhaps, the most universally liked of all the world's top racing drivers, as well as highly respected for his engineering abilities, and his death had come as a tremendous blow to the grand prix world. The fact that there was about to be a race on what was generally considered by grand prix drivers to be one of the most hazardous circuits in the world, especially in the wet, served to heighten the general feeling of depression.

For the many users of the Ford engine there was an additional problem. For several weeks there had been a considerable shortage of raceworthy engines as a result of a crankshaft breakage problem caused by a machining error by an outside supplier to Cosworth.

This had served to increase tremendously the already heavy workload being undertaken at Northampton, and despite some prodigious efforts by the Cosworth staff the temporary shortage of healthy engines was now critical, one or two teams no longer having any spares, and John Surtees for one having to withdraw from the Belgian race for the lack of an engine. In the circumstances, therefore, it was not surprising that one or two teams elected to complete the minimum of practice.

Ken Tyrrell's team was reduced to one car, but in this case it was because Servoz-Gavin had suddenly announced his decision to give up racing and no substitute had yet been named. But with Jackie Stewart in a relatively unchallenged pole position Tyrrell was still hoping to avenge the defeat of 1968 when fuel shortage had robbed Stewart of a comfortable victory.

Rindt and Amon shared the front row of the starting grid with Stewart, the Austrian still using his Lotus 49C for the race in preference to the wedge-shaped Lotus 72. Rindt had first driven the new car in Spain, where it had revealed a number of handling deficiencies which had subsequently been confirmed at Silverstone, and although some extensive detail redesigning had since been carried out, mainly involving strengthening the monocoque structure and removing some of the anti-dive from the suspension, Rindt decided that for one more race he would opt for reliability rather than for pace. In the event he was to achieve neither, and he retired with a sick engine before half-distance.

Chris Amon was still hungry for that first championship race win, and he swept straight into the lead at the start, and although he was repassed by Stewart on the second lap he regained first place very quickly and led for two further laps. But already Pedro Rodriguez, who had started his BRM from the third row of the grid, was beginning to make a big challenge on a circuit where a few weeks earlier he had put up a prodigious performance in a Porsche sports car.

He had revealed something of the BRM's improved performance at Monaco by climbing through to sixth place after an early pit stop, and now he was demonstrating that the car had a lot of speed on a fast circuit. By the fifth lap he had put the P153 through into first place, and although Amon tacked on behind and remained glued to the BRM's slipstream throughout the race he was just unable to find a way past, and he took the chequered flag just over a second behind the Mexican.

Stewart's race had ended at half-distance with engine trouble, and Jack Brabham, who had also been challenging hard in the early stages, retired soon afterwards with a loose clutch ring gear, and this allowed the 12-cylinder cars to take over most of the leading places. The two Matra-Simcas of Beltoise and Pescarolo ran nose-to-tail in third and fourth places for many laps until Pescarolo stopped for more fuel, although Jacky Ickx had to make a pit stop when lying third to cure a fuel leak in the cockpit and to pour water over his soaked overalls which were aggravating his already painful burns. However, Ferrari had started a second car in a grand prix for the first time that year, and Giunti did a worthy job to bring it home in fourth place at the finish, ahead of Stommelen and Pescarolo.

Whilst the BRM victory – their first with a 3 litre car – was a great surprise, the more so because the team had had an abundance of problems during practice and Oliver's car had run into engine trouble early in the race, the Belgian Grand Prix had served to demonstrate the growing competitiveness of the 12-cylinder cars, which suggested that an exciting power struggle could develop during the remainder of the season – just the sort of entertainment which spectators like to watch.

Meanwhile, more development tests with the Lotus 72 had given Jochen Rindt renewed confidence in the new car, and he approached the Dutch Grand Prix at Zandvoort with considerable enthusiasm and a degree of confidence which was to prove completely justified. After setting the fastest time in practice to line up alongside Stewart and Ickx, he took over the lead from the Belgian on the third lap and thereafter was virtually unchallenged. It was a most convincing victory for both car and driver, but it brought Rindt little joy, for a little over a quarter of the way through he had seen his close friend Piers Courage die in the blazing wreckage of his de Tomaso, which had crashed into a bank at high speed. Zandvoort, normally such a cheerful place on the evening after the race, lacked its customary carnival atmosphere that night as motor racing mourned its second major tragedy in less than three weeks.

Even without the de Tomaso accident to distract them, most drivers would have found the Dutch Grand Prix a hard race, for the track became extremely slippery in places, making it very difficult for drivers to pass each other. Oliver and Rodriguez went well during the early laps of the race, but the BRMs were out of the running well before half-distance, Oliver suffering more engine trouble and Rodriguez having to stop twice to fix loose bodywork (he was using a spare car, having written-off his own a few days earlier in a high-speed accident during practice).

Ferrari were again running two cars, this time with Regazzoni in the second 312B, and both finished, Ickx in third place after a quick pit stop and Regazzoni fourth, while Beltoise again revealed that the Matra-Simca had reliability, if not quite sufficient speed, by taking fifth place ahead of Surtees in his McLaren M7C. The McLaren works team were back in action, with Dan Gurney accepting an invitation to rejoin the team in place of Bruce, while Peter Gethin had been brought in to deputize for Denny Hulme, whose hands had been badly burned in a testing accident at Indianapolis. But Gurney's race was over within three laps with broken timing gears, while Gethin's Dutch Grand Prix debut came to an abrupt halt when he left the track on a particularly slippery corner, and badly creased the monocoque chassis. The other newcomer in the race, Francois Cevert, who had been given the second car in Ken Tyrrell's team in place of Servoz-Gavin, drove strictly to orders and at least had the satisfaction of travelling further than either of the works March drivers before his car broke down. For everyone, though, this had been a desperately unhappy race.

The growing challenge by the 12 cylinder cars was revealed clearly during practice for the French Grand Prix at Clermont-Ferrand two weeks later when the struggle for fastest lap developed into a private duel between Ickx in his Ferrari and Beltoise in his Matra-Simca, Ickx eventually securing pole position by a fraction of a second.

Beltoise, of course, was desperately anxious to follow up the previous year's Matra one-two on the Clermont circuit with an all-French victory this time, and for much of the race it looked as though he would succeed. Ickx had shot into the lead at the start, but even before the race had begun it was known that all was not well with the Ferrari's engine, and Beltoise had been content to consolidate his second place in the hope that the Ferrari would run into trouble.

Sure enough, on the 14th of the 38 laps, Ickx's car began to slow as his engine gave all the symptoms of valve trouble. The next lap Beltoise went triumphantly through into the lead as thousands of Frenchmen roared their approval. With Ickx dropping back and soon to retire, Jochen Rindt became Beltoise's main threat, but as the gap between the two cars was over 10 seconds and widening every lap his first place seemed to be in no danger.

For most of the race Rindt had been battling closely with Amon, and the fact that second place was now in dispute served to intensify the struggle between them. But Rindt was endeavouring to pace himself very carefully. He needed no reminder that nausea had caused him to drop out of the race the previous year, and although he no longer felt any after-effects of his 1969 Spanish Grand Prix accident, the Clermont-Ferrand circuit remained a strong challenge to his personal metabolism. Also, he was still suffering slightly as a result of an early practice incident when a stone thrown up by the car he was following had struck him in the face and caused him to rush away from the circuit to have the wound stitched. He was in no mood, therefore, to over-tax himself.

Beltoise continued to extend his lead, and with the car sounding in excellent health it looked as though the race result was a foregone conclusion. But suddenly the Matra-Simca's lap times started to lengthen dramatically, and soon it was obvious that Beltoise had a handling problem, the reason for which became clear when he shot into his pit on lap 27 with a punctured rear tyre. Both rear wheels were changed as a precaution, and Beltoise went back into the race, but he was now down in tenth place, and when he had to make a second stop shortly before the finish to take on more fuel he decided it was not worth continuing.

Rindt, of course, had swept into the lead immediately Beltoise had slowed down, still with Amon a few seconds behind him, and the Lotus and the March circulated in fairly close company until the end of the race, the gap between them as they crossed the line being less than eight seconds.

An even closer battle was fought by Jack Brabham and Denny Hulme, making his racing comeback with his hands still painfully tender, and only a fifth of a second separated the Brabham and the McLaren at the finish, while Dan Gurney's challenge to Pescarolo's fifth place failed by only two fifths of a second. It had been an interesting and exciting race, and although there was universal sympathy for Beltoise, Rindt was naturally delighted to pick another victory out of the bag because it meant that now he had 27 world championship points, eight more than Brabham and Stewart, whose early challenge at Clermont-Ferrand had disappeared when he stopped to change another faulty spark box on the Tyrrell Matra.

1970 was to produce several unusually exciting Formula 1 races, but none was to match the British Grand Prix, the sheer drama of which surpassed even that which had been seen at Monaco two months earlier. Two days of practice at Brands Hatch had indicated that the main action would be provided by Rindt and Brabham, who had shared the fastest time, with Ickx, having secured the third place on the front row, providing the main challenge to the Ford-powered cars. It turned out to be an accurate prediction of the structure of the race, but gave no hint of the electric atmosphere which was to be generated both during the grand prix and for several hours afterwards.

Jack Brabham out-accelerated Rindt and Ickx from the start line, but Ickx retaliated immediately and managed to find a way past on the opening lap before setting about consolidating a lead. But his first place was to be short-lived, and he retired after only six laps with a broken differential. Almost simultaneously, Rindt had put in a strong challenge to Brabham's second place so that in one passing manoeuvre he suddenly found himself elevated from third place to first, a very rare achievement at Brands Hatch.

Brabham, however, had no intention of allowing Rindt to get away from him, and he settled down close behind, prepared to bide his time before launching a counter-attack. Once again Jackie Oliver was putting up a good performance at Brands Hatch, and was holding his BRM in a solid third place ahead of Hulme's McLaren and Stewart's March, which for the second time that year was not showing up too well on the challenging circuit. But Brabham and Rindt were opening the gap ahead of all the other cars lap by lap until the Grand Prix had become a two-car race.

It was noticeable that whenever Brabham lost some ground through being delayed by slower traffic he had little difficulty in closing up again on the Lotus once he had passed the back-marker. There was little doubt, therefore, which was the faster car on the day, and the only question remaining was when Brabham would launch his attack and move back into the lead. He timed his manoeuvre with 11 of the 80 laps to go, seizing the opportunity to shoot past as Rindt missed a gear and hesitated momentarily.

Once in front, Brabham set about consolidating his lead because he wanted no late Monaco-like challenge from the Lotus driver to distract him on the final lap. There seemed little chance of this happening, and as Brabham was shown the 'one lap to go' signal as he passed his pit for the last time he was holding a useful lead of 14 seconds. But half-way round that final lap the Brabham started to falter, and coming out of the last sharp left-hand corner, Stirling's Bend, the engine cut out altogether; it was out of fuel! An astonished Jochen Rindt passed the sticken Brabham as it freewheeled towards Clearways and the finishing straight, the Lotus leader taking the chequered flag for the third Formula 1 race in succession. A black-faced Jack Brabham followed him across the line 33 seconds later, climbed out and walked silently back to his pit as Hulme's McLaren just beat Regazzoni's Ferrari into third place.

But the drama was not yet over The scrutineers examined the winning Lotus after the race and questioned the height of the wing, which they suggested was above the maximum allowed by the regulations, and more than an hour after the race had finished it was announced that the Lotus had been disqualified. So Brabham was the winner after all! Or was he? Colin Chapman had been drinking victory champagne when someone informed him of the disqualification, and he immediately put in an appeal. The Lotus wing was stripped down, rebuilt, stripped again, and reassembled in different positions before the officials were satisfied that it did, after all, just meet the regulations. All the drama and uncertainty, therefore, had been for nothing, and Rindt could satisfy himself that nine more points were in the bag, while Brabham was no longer in any doubt as to which race he would name the unluckiest of his career should he ever decide to write his memoirs.

The second half of the 1970 world championship season began with the German Grand Prix, and this meant a visit to unfamiliar territory for most people as the race organizers had been unable to meet the demands of the Grand Prix Drivers' Association in respect of safety measures at the race's traditional home, the

Nurburgring. Hockenheim, a short distance from the Frankfurt-Stuttgart autobahn, was chosen as the alternative venue, and although the circuit, which forever will be linked with Jim Clark's death, had been severely criticized in the past both for its dangers and for its concept of very long straights and a bunch of extremely slow corners, it provided a magnificent setting for a wonderfully close-fought and safe Grand Prix.

An 'S' bend had been inserted at two of the fastest points on the course to help reduce the speed of cars along the straights, but nevertheless it was anticipated that horsepower would still be all-important at Hockenheim, and that practice times might be more than usually revealing.

Regazzoni, whose performance in the Ferrari 312B at Zandvoort and Brands Hatch had revealed his rapidly growing maturity as a racing driver after his somewhat bizzare early career, was the first to break the two-minute barrier on the modified circuit, although both Ickx and Rindt were to beat his time before the end of practice. In the final line-up it was interesting to find one Ford-engined car and one 12-cylinder car on each of the first four rows, Rindt, Siffert, Amon and Stewart (a Lotus and three Marches) being placed alongside Ickx, Regazzoni, Pescarolo and Rodriguez (two Ferraris, a Matra-Simca and a BRM), while all but two of the remaining 13 cars were Ford-powered.

Even with its modifications, Hockenheim remained a circuit where a slipstream tow could be invaluable, and therefore it was vitally important at the start of the race to become part of the leading bunch of cars. Understandably, Ickx, Rindt, Regazzoni and Siffert set the pace from the first two rows, while Amon did a magnificent job during the first few laps to break loose from the main bunch of cars behind and tag on to the end of the leading quartet, later replacing Siffert when he dropped back with a misfiring engine.

For more than half the race the two Ferraris, the Lotus and the March raced wheel-to-wheel in a tightly knit bunch, sometimes changing places several times during a lap. It was motor racing at its best for on any lap it was anybody's guess who would be in front as the cars rushed into sight of the vast filled-to-capacity grandstands, even though Rindt and Ickx were doing their best to monopolize first place.

Ickx had the advantage of an ally in Regazzoni, and the two drivers worked extremely well together as a team, Regazzoni doing well on several occasions to shield Ickx from Rindt and Amon and so allow him to make a break. But always Rindt fought back, and with 30 of the 50 laps completed Ickx suddenly found himself without an aide when Regazzoni's car suddenly seized its engine and slid to a halt.

Only three cars now remained to fight for the lead, and when Amon started to drop back as his engine became tired it was left to Ickx and Rindt to settle the matter by themselves. With five laps to go the Ferrari was fractionally in front; four laps to go and the Lotus was ahead; three laps and the Ferrari was back in front again. But for the final two laps Rindt was able to keep the Lotus ahead as the two cars crossed the start-finish line. He had won his fourth grand prix in succession and his fifth of the year, and although sadly he was never to know it the 45 points he had amassed were to be sufficient to earn him the world championship title which had been his life's ambition.

Denny Hulme and Emerson Fittipaldi were the only other drivers to complete the full race distance, Fittipaldi, who had had his first Formula 1 race in a works Lotus 49C at Brands Hatch, putting on such a fine display in this relatively old car that it was to earn him a permanent position with Gold Leaf Team Lotus.

Another newcomer at Brands Hatch had been John Surtees' new grand prix car, the TS7, which had revealed an impressive chassis performance before retiring with failing oil pressure. At Hockenheim as at Brands Hatch Surtees had been forced to start his car near the back of the grid, but he did a magnificent job in working his way through the field and was running in third place until four laps from the end when again he ran into engine trouble. To retire so late in the race was a bitter disappointment, but at least he had the satisfaction of proving without any doubt that, despite his years of frustration in grand prix racing, he was still one of the most formidable drivers.

Unfamiliar territory was also provided for the next round of the world championship series, the Austrian Grand Prix having been readmitted to the schedule now that the organizing club had a worthy new circuit in the Osterreichring, close by the race's former airfield home at Zeltweg.

The circuit seemed to suit the Ferraris extremely well, and the Italian team decided to enter three cars as a dress rehearsal for their all-out effort to win the following month's race at Monza. It looked as though Regazzoni and Ickx would end up with the two fastest practice times and that Giunti in the third car would be only fractionally slower, but a superhuman effort by Rindt in front of his home crowd earned him the coveted pole position, while Stewart managed to displace Giunti from the second row of the grid.

It was Regazzoni who made the initial running from the front row and Ickx managed to slip past Rindt into second place as the large 24-car field left the starting area. Unfortunately, the engine in Cevert's March blew up on the first lap and the resulting pool of oil made the track treacherously slippery at one point, and on his third lap Rindt eased right back when he saw the oil flag being waved furiously, and dropped behind Beltoise, Giunti and Amon to sixth place.

This was a big blow, because it meant that the Ferraris could now make a clear break, but Jochen worked tremendously hard to make up the deficit, repassing Amon on his sixth lap and moving ahead of Giunti into fourth place five laps later. Then he began the job of hauling-in Beltoise, and was just getting on terms with the Frenchman when suddenly his engine gave an ominous rattle. He switched it off and parked his car beyond the start line; Jochen Rindt's last grand prix was over.

Jacky Ickx, who had taken over the race lead on the second lap, once again had Regazzoni protecting his tail, although this time there was only one car, Beltoise's Matra-Simca, offering a close challenge as Giunti was doing a useful job for his team-mates in keeping the other cars at bay further behind.

Beltoise drove extremely well to keep in contact with the leaders, although there never seemed to be much chance that he would get past them, any hope that he might have had of doing so being dashed just five laps from the end when once again the Matra-Simca ran out of fuel. With Jackie Stewart dropping out early on with a split fuel line, Brabham stopping to replace a holed radiator, and Hulme retiring with engine trouble, it was left to Rolf Stommelen to take up the main Ford challenge. The Brabham driver put in one of his best performances of the season and was the only other person to remain on the same lap as Ickx and Regazzoni as they crossed the winning line nose-to-tail, having finally scored the race victory for which the Ferrari team had been striving for so long. With Rodriguez and Oliver also finishing in line ahead in fourth and fifth places, and Beltoise beating Giunti, who had been delayed by a puncture, into sixth place, it was an impressive result for the 12-cylinder cars, and an indication that in its fourth year of active racing life the Ford DFV engine had at last found some worthy rivals, which would serve to heighten interest in the remaining 1970 races, as well as encourage Keith Duckworth to extract a little more power from the DFV for the 1971 season.

One week after the Austrian Grand Prix, the International Gold Cup race at Oulton Park was opened to both Formula 1 and Formula 5000 cars, of which, understandably, the grand prix cars and their drivers were to emerge clear victors. The race was held in two parts, and was to give John Surtees his first victory with his Ford-powered TS7, but not before he had endured a few anxious minutes during the closing laps of part two listening to a transmission vibration. Surtees had won the first part of the race by 6·6 seconds over Jackie Oliver's BRM P153, with Jochen Rindt's Lotus 72 a further 6·2 seconds behind in third place.

Rindt's car had not been handling well in the first race, but adjustments to the wings during the interval made a big improvement, and he shot into the lead of the second race while Surtees spent several laps finding a way past Oliver. But once John was through his only worry was to keep within 12·8 seconds of Rindt until the end of the race, which he managed to do with less than three seconds to spare. After crossing the finishing line the Austrian parked his car at the side of the circuit, then sprinted across to a waiting plane which was standing by to rush him to an urgent appointment on the Continent. The scene was shown on television screens throughout Britain that afternoon; it was the last view most people were to have of the man destined to be the new world champion.

Jochen Rindt's death while practising for the Italian Grand Prix at Monza less than two weeks later was the final bitter blow in a desperately tragic year for grand prix racing. It occurred at a time when Rindt was content, for the time being, to collect just a few more points in order to put his claim to the championship beyond doubt, whereas earlier in the year he had been driving as hard as only he knew how to win race after race. But whether you are driving on the limit or merely close to it, a high-speed accident can have terrible consequences, and this time Jochen Rindt was to be robbed of the good fortune which had allowed him to escape so lightly from his accident the previous year in Barcelona. His Lotus crashed with tremendous impact into the guardrail entering the *Curva Parabolica*, and he died almost immediately.

Understandably, Gold Leaf Team Lotus withdrew from the race, and Rob Walker's new Lotus 72 which Graham Hill was to have driven was also put back into the transporter while the remaining drivers did their best to concentrate on practising and setting-up their cars for the following day's race.

This time Stewart's March was the only Ford-powered car on the first three rows of the starting grid, surrounded by three Ferraris and two BRMs, which was proof enough that Jackie, saddened by the loss of his close friend and neighbour and weary through loss of sleep, would have a hard race ahead of him.

A few weeks earlier, Ken Tyrrell had caused a considerable surprise by unveiling a brand new car bearing his own name which had been designed for him by Derek Gardner, who previously had been employed by Harry Ferguson Research Limited and, as such, had been involved with the Matra MS84 four-wheel-drive car. The new car, which it was hoped would be more competitive than the March 701, not surprisingly had a considerable amount in common with the 1969 Matra MS80, which Stewart had always considered to be the best-handling single-seater he had ever driven. Some of the Tyrrell 001's potential had been revealed during the International Gold Cup race at Oulton Park, where Jackie had set a new lap record with apparent ease before retiring, but although the car had been brought to Monza it was decided not to race it as there had been insufficient time to sort out various minor problems with it in the aftermath of Jochen Rindt's accident. In any case, the March had gone extremely well in practice, to give Jackie the fourth fastest time on the grid.

When the flag was dropped Ickx shot straight into the lead, and as Rodriguez, who had started beside him in the BRM, was delayed fractionally by wheelspin Stewart only just failed to accelerate into second place as the cars streamed on to the narrow part of the main straight. Stewart was still third at the end of the first hectic lap, and shortly afterwards was passed by Regazzoni. But at least he was part of the leading bunch, which was all that mattered this early in the race.

Although Ickx had completed his first lap with a useful lead he was unable to shake the others out of his slipstream, and during the first ten laps Rodriguez, Stewart and Regazzoni also took turns to lead the race. But Rodriguez ran into engine trouble very soon afterwards, to be followed almost immediately by Pescarolo and Giunti, while Ickx's car retired with transmission trouble well before half-distance. This meant that Regazzoni was the lone survivor of the Ferrari trio, a heavy responsibility in view of the fact that he had been unable to shake off a formidable bunch of seven cars comprising Stewart's March, Oliver's BRM, Hulme's McLaren, the Brabhams of Jack himself and Stommelen, Beltoise's Matra-Simca and Cevert's March.

The first of these to drop out was Brabham's BT33, which had been suffering from intermittent fuel starvation and which suddenly cut dead half-way through the 180-degrees *Curva Parabolica*. Starved of power, the car suddenly ran wide and slid into the barrier, ripping off a wheel, and the incident also cost Cevert his chance of a high placing at the finish because in slowing to avoid Brabham's sliding car he lost his tow. Soon afterwards, Oliver's BRM, which had been going very well indeed, and leading the race for a number of laps, suddenly broke its engine and was abandoned just beyond the finishing line, so that now only five cars remained in the leading bunch.

Regazzoni kept a cool head under the pressure, and after waiting until ten laps from the end before making his big effort to break clear, he succeeded and pulled out a vital three seconds over Stewart, which was sufficient to break the tow. By the end of the race he had extended his lead to nearly six seconds, and while the Italians went wild with delight Stewart, Beltoise, Hulme and Stommelen set about dividing up the next four places between them. This, in fact, was their finishing order after the inevitable frantic jockeying of positions on the 68th and final lap, less than seven-tenths of a second covering the four of them as they passed the chequered flag. It was an exciting ending to a fine race, but the numbing effect of Jochen Rindt's death was to take a long time to wear off.

With two grand prix successes in succession, Ferrari spirits were very high as all the European-based teams made their annual Transatlantic journey for the final three races of the championship series. But although the Ford DFV engine had been beaten it was far from eclipsed, as Jackie Stewart demonstrated so convincingly during practice for the Canadian Grand Prix, which had returned to the St. Jovite circuit in Quebec.

Once again Stewart tried both the March 701 and the Tyrrell 001, and after overcoming a problem with loosening wheels on his new car he used it to set the fastest lap of the day to secure pole position on the grid, and decided to race it even though its engine did not feel quite as free as the DFV installed in his March.

Stewart powered his way into an immediate lead at the start of the race, outsprinting Ickx who had begun the race alongside him, while Rodriguez shot up from the fourth row with his BRM to slide into third place. There was considerable consternation in the Ferrari pit as lap after lap Stewart proceeded to extend his lead until he had more than 18 seconds in hand over Ickx with 30 of the scheduled 90 laps completed, while Regazzoni, in third place, was a further 21 seconds behind and being hard-pressed by Amon. But going into his 32nd lap Stewart found his car weaving badly, and after a few anxious moments he brought it to rest with a broken front hub as a considerably relieved Ickx went through into first place.

From then on it was Ferrari first and second all the way, while Amon, who had been unable to keep pace with Regazzoni during the second half of the race, had to resist a very strong challenge from Cevert in Ken Tyrrell's March until the Frenchman had the misfortune to break a shock absorber shortly before the finish. Rodriguez's BRM ran reliably throughout the race until it ran out of fuel near the end, and the Mexican was fortunate to retain his fourth place after his quick pit stop, crossing the line a short distance from Surtees, who had scored his first world championship points with his new car.

By winning the Canadian Grand Prix Jacky Ickx had retained a slim chance of overtaking Jochen Rindt's total of 45 world championship points in the two remaining races, but much as he wanted to win every race he entered he expressed the hope that Jochen's right to the title would remain undisputed as 1970 had been so convincingly his year.

Jacky's honourable and unselfish wishes were to be met two weeks later at Watkins Glen, where a pit stop during the United States Grand Prix to replace a broken fuel breather shortly after half-distance dropped him from second place to twelfth, from which a fighting comeback was to elevate him to a gallant fourth place at the finish. Ickx's points total after Watkins Glen was only 31, which meant that even if he won the Mexican Grand Prix there would still be a margin of five points in favour of Rindt.

Jacky Ickx had placed his car in pole position during practice for the United States Grand Prix, but with Stewart also on the front row with the Tyrrell he was unlikely to be given an easy race. In fact, all he saw of Stewart's car was its rapidly disappearing rear end, for Jackie was soon repeating his St. Jovite performance of rushing off and hiding from everyone else in the race.

Rodriguez, who had started his BRM immediately behind Stewart's Tyrrell, was able to out accelerate Ickx on the initial uphill sprint beyond the first corner, and this enabled Stewart to make an early break from the Ferrari, which was held up behind the BRM for the first 15 laps. At that point both Ickx and Regazzoni managed to scramble past Rodriguez, but by this time Stewart had built himself an impressive lead and was dictating the race completely.

Regazzoni's hope of a high placing was dashed first when he punctured a tyre and later when he made two more pit stops with a misfiring engine, and by the half-way point Stewart was so far ahead that he had lapped all but Ickx and Rodriguez in second and third places. Gold Leaf Team Lotus had made a return to the grand prix scene with a new team of drivers, Emerson Fittipaldi having been elevated to lead the team – this was his first race with the Lotus 72, which he had been scheduled to drive at Monza – and Reine Wisell of Sweden having been signed in place of John Miles, who had dropped out of the Formula 1 team. The two newcomers were driving extremely well in their first grand prix together, and were lying fourth and fifth at the half-way point, ahead of Derek Bell, who was having his first drive in a Surtees TS7.

Once Ickx had stopped with his fuel-breather trouble only Rodriguez remained on the same lap as Stewart, but there seemed no possibility of the Mexican catching him. But suddenly the Tyrrell started to trail oil smoke, and gradually the smoke cloud became more intense as the engine oil was consumed more and more rapidly. By the end of the 82nd lap, with 26 still to run, the engine suddenly sounded very rough, and Stewart knew that his race was over, while a delighted Rodriguez moved into the lead and suddenly saw his chance of a second grand prix victory with the BRM.

But with 100 laps completed and the race seemingly won the thirsty BRM ran short of fuel, and despite the lightning-quick work of the BRM mechanics and a most spectacular exit from the pit road by Rodriguez the Mexican was now too far behind Fittipaldi to stand any chance of catching him in the eight remaining laps; his own first place had been inherited by good fortune, but now misfortune had knocked him back to second place again.

For Gold Leaf Team Lotus it was an unbelievable result, but a magnificent comeback from their Monza tragedy, for the victory of Fittipaldi, backed up by Wisell's third place, not only ensured that Jochen Rindt's world championship was beyond the reach of anyone else, it also confirmed the Lotus-Ford as the championship car for 1970. The young Brazilian, who a year previously had been just another Formula Ford driver, and had since taken part in only four world championship races, had won the richest of them all. Colin Chapman, for whom the previous few weeks had been intensely difficult, was completely overcome during the celebrations immediately after the race when he realized the full meaning of Emerson Fittipaldi's victory.

After such a remarkable race, and with both world championships decided, the Mexican Grand Prix, the final race of the series, had been provided with all the ingredients of an anti-climax, but there was to be no shortage of headlines as journalists prepared their stories afterwards.

They were concerned not so much by the fact that Ickx and Regazzoni had scored another one-two victory for Ferrari after Jackie Stewart's efforts to remain with them had ended when he had suffered a loose steering column bush and subsequently had hit a stray dog and damaged his suspension, as by the fact that the race had taken place at all. For hours beforehand spectators had poured on to the circuit, completely disregarding the efforts of the police and officials to marshal them back behind the safety barriers. Even Stewart and Rodriguez were prevailed upon by the race organizers to plead with the crowd for their own safety, but all in vain, and when the start was delayed tempers became so frayed that race officials had to warn entrants and drivers that there could be a major riot, with perhaps disastrous consequences, if they refused to start the race. It would be safer for all of them, they were told, to race between the lines of spectators sprawled on the grass beside the track, or even walking across it, than not to race at all!

After an exploratory lap the drivers agreed amongst themselves to take things steadily during the first few laps of the race while they sized up the situation, and then to be prepared to abandon the race if conditions deteriorated. But a few deliberate weaving manoeuvres by one or two of them during the opening lap had a satisfactory sobering effect on some of the more unruly elements, and conditions remained just tolerable throughout much of the race. But during the closing laps hordes of spectators poised themselves to invade the track and there would have been a major catastrophe had not the race organizers decided wisely to show the chequered flag at the final corner instead of in front of the pits and main grandstand. But their complete failure to maintain crowd control and to provide conditions in which drivers could race in reasonable safety was to lose them their world championship status for the 1971 season.

The Grand Prix, perhaps because of its crowd problems, was a rather processional race, with Jack Brabham providing the main opposition to the two Ferraris following Stewart's pit stop, and seeming certain to gain a third place in his very last grand prix – he had announced his retirement immediately before the race. But the ill fortune which had dogged the amiable Australian throughout much of his 23rd and final year as a racing driver remained with him to the end, and he retired with an engine problem 13 unlucky laps from the end. Hulme and Amon, who had been chasing Brabham hard, were left to follow Regazzoni home in third and fourth places, a short distance from Beltoise in the Matra-Simca, while the last point of the 1970 season was earned by the local hero Rodriguez, who, like Oliver, had found his BRM considerably affected by the thin air of Mexico City.

The hectic, sometimes exciting, sometimes sad, 1970 grand prix season was at an end. The battle had been close-fought, and Jochen Rindt and the Lotus-Ford had emerged the worthy winners. It had been the fourth season of racing for the Ford DFV engine and the first time that it had been challenged very seriously. That it had managed, if only by a relatively narrow margin, to fight off that challenge, despite a serious maintenance problem at the height of the season, and despite the fact that it was still substantially the same engine as had appeared at Zandvoort in 1967 – no serious effort having yet been made to extract more power from it – was perhaps the best possible testimony to the soundness of Keith Duckworth's original design.

Those who were suggesting that its days were over, and that the 12-cylinder power units would be leaving the DFV far behind in 1971, were to be confounded by the facts, and it was significant that those who had most cause to evaluate its status most carefully, the entrants and the drivers, almost without exception shared Duckworth's conviction that there was considerably more performance still to be tapped from the lightweight, compact V8.

The queue for DFV engines remained as long at the end of 1970 as at any time in the engine's life, for Brabham, Lotus, McLaren, March, Surtees and Tyrrell were all intent on relying on Ford power again for the sixth year under the 3 litre formula, and had already designed their new cars around it.

Keith Duckworth and his colleagues set to work at Northampton on their first serious attempt to extract more power from the DFV, and at the same time they announced a reorganization of the company's maintenance service which would benefit not only their customers but also themselves. Whilst they would remain entirely responsible for the maintenance and overhaul of all their latest-specification engines, they would be supplying full servicing and overhaul facilities to selected outside companies, who would be entrusted with the job of maintaining the earlier DFV engines for those customers who were unable or unwilling to operate their own engine service department.

It had been a near-miracle that Cosworth Engineering had been able to carry out this work single-handed for so long, and perhaps it was fortunate that for three seasons the DFV engine had proved so superior to its opposition that the minimum of Cosworth's resources were required for development work, so that the maximum could be spared for maintenance, repair and routine overhaul. But now that the opposition had contrived to close the gap it was time for Cosworth's resources to be spread more evenly between maintenance and development, if only to protect the long-term interests of their many customers.

Having seen the very existence of long-established and well-organized racing teams threatened by the lack of a competitive power unit, and having so successfully taken up the challenge to plug the gap, Keith Duckworth, his co-directors and his staff, were now determined to ensure that such a situation as occurred in 1961 and again in 1966 would never be allowed to happen again.

The DFV engine already had contributed to a hat-trick of three world championships for both drivers and constructors, and had powered the winning car across the line at the end of no less than 34 grands prix, but this was not the conclusion of the DFV story. It was just the end of the first chapter.

11

Improving the breed

DESPITE KEITH DUCKWORTH'S oft-quoted remark about development only being necessary because of the ignorance of designers, the intensity of competition and the speed of progress in motor racing is such that a continuous programme of product improvement has to take place with all the major components of a racing car, and perhaps most of all with its engine.

Notwithstanding the remarkable achievement of the DFV on its first appearance in the Dutch Grand Prix in 1967, it was clear to Duckworth when he examined both Graham Hill's damaged and Jim Clark's seemingly healthy engine after the race that all was not entirely well with either power unit. It was at this moment that the DFV's development programme, which has continued non-stop ever since, was set in motion, and much of the credit for the engine's overwhelmingly successful career during the subsequent four years must go to Cosworth Engineering for the painstaking manner in which they carried out this aspect of their work.

Watching the DFV engine power car after car to race victory after race victory, sometimes to be followed home by similar engines in the next five cars, it had been all too easy to assume that once the design had proved itself it had become plain-sailing for Cosworth, whose job was simply to put a few more engines down the production line from time to time, and to service and occasionally repair those already in use. But had they adopted such a policy the DFV story might have been very different.

Although fundamentally the DFV engines of 1967 and 1971 were the same, they differed in many details of their design, for at some time during the intervening period scarcely a component escaped the attention of Keith Duckworth and his colleagues as they went about their task of improving the breed.

As in the earlier chapter on the design and philosophy which went into the DFV engine, Duckworth's own words are used to describe the subsequent development of the engine as he spoke them to David Phipps.

'When the engine was first run it became apparent that the oil drainage from the heads and the removal of the blow-by were grossly ill-considered. As a result we put in a lot of overtime trying to devise a system that would stop the engine filling with oil, and we ended up with a very crude arrangement of external pipes that stayed with us for the whole of '67. As soon as we got this working the engine gave about 408 horsepower, so we deemed we had enough power to go motor racing.

In the first race at Zandvoort Jimmy's car had hub trouble and he didn't get much practice, but Graham started from pole position and was leading the race when he was forced to retire with a timing gear failure. Jimmy went on to win, but when we stripped his engine we found a tooth missing from one of the timing gears. We had already had a gear failure on the test bed, but at that stage we felt that the gears themselves were not as good as they might have been, and that the backlash was much too high.

We were very pleased that Graham had been able to lead the race so easily and that Jim went on to win in a car that had scarcely run. Apart from the gears there did not seem to be much wrong with the engines, so we rebuilt them and then built a few more. At Spa we had a plug problem, and we had the odd nose fall off a camshaft, while the French race would have been all right had it not been for the gearboxes failing. After that the gearbox casing was strengthened, but it was still something of a disadvantage to be stuck with a fixed set of ratios and to have to change the whole transmission if it was necessary to alter the final drive – particularly as people with Hewland gearboxes were able to change individual ratios and get the right gearing for particular corners.

While designing the engine we had had talks with Lotus about the best way of fitting it in the car, and in particular about making reasonable use of it as a structural member. We had said we would produce mountings on the bottom of the sump and the ends of the rocker covers, and had thus defined the shape of the back bulkhead of the monocoque. The bottom mounting bracket was intended to be a weak link, so that if there was a shunt the aluminium adaptor between the chassis and the sump would fail, rather than the sump or the crankcase. This was a very light-looking component, and we spent the first year crack-testing it after every race, but we have never had a failure or even a crack in one – much to our surprise.

At Silverstone the nose fell off a camshaft on Graham's car. This was after Graham had had a suspension failure in practice and the car had been rebuilt overnight and had then had a suspension bolt come loose while leading the race; it was all a bit dramatic. Jimmy was faster in practice and he won quite easily, but he had a problem with dirt in the fuel system which was blocking the filters and causing loss of fuel pressure. We had this trouble several times in 1967, and we thought it was probably due to material coming away from the inside of the tanks. In the end we built some larger filter housings and used a much bigger filter; this seemed to overcome the problem in the main, although the fuel pump is very sensitive to dirt and has to be very accurately built and fitted to guarantee an adequate fuel supply. Seizures of the pump because of dirt are still fairly common on certain cars.

At the Nurburgring both cars dropped out with suspension trouble, and in Canada Clark's ignition packed up while he was in the lead. Afterwards we found that this was due simply to an electrical plug getting damp; we had made allowances for wet weather by covering the whole ignition system, but the damp got into the plug which connected it to the engine.

At Monza we had an oil pressure problem. It was very hot and we couldn't maintain satisfactory oil pressure on the synthetic oil we were using at that time. As a result of this we designed new oil pumps with a larger capacity, and these were fitted for the Mexican race; until they were ready we used a higher viscosity oil and this seemed to be quite satisfactory. Graham had a crankshaft break at Monza while he was leading by over a minute, and Jimmy ran out of fuel after making up over a lap following a puncture. Graham's crankshaft failed because the steel from which the cranks had been made was fairly dirty; as a result of this we changed to vacuum remelted steel, but we had to wait six months before we could get any.

Graham was supposed to win at Watkins Glen – he and Jimmy had tossed for it – but his clutch ring disintegrated and Jimmy had to pass him to prevent Amon catching both of them. Then Jimmy's rear suspension broke and he did the last three laps with rather a lot of negative camber. Graham finished just a few seconds behind him; it was our first one-two. Jimmy won again in Mexico, but Graham dropped out with a broken drive-shaft joint.

By the end of 1967 we had decided that we didn't like the look of the external oil pipes – particularly as we were still having a certain amount of scavenge trouble even with them on. In 1968 we introduced an air pump, which was supposed to transfer air and blow-by from the bottom half of the engine to the top half. Unfortunately a lot of oil went up with the air, so this didn't work out well. Eventually we decided to use the air pump as an extra oil pump; instead of transferring air from the bottom of the engine to the top we sent the total output of the scavenge pumps and the air pump back to the tank, and ran the whole engine at a depression. In this way we removed more than the total blow-by of the engine as well as the oil, and allowed air to enter the engine through what was formerly a breather.

This posed quite a problem for most of the car manufacturers. There was so much air and oil going back to the tank that their separation systems were inadequate. Several people blew all their oil out through their tank breathers and ran out of oil, causing a number of bearing failures. As a result I thought it would be sensible to try to make a separation system on the engine. Unfortunately, although the device which I had designed to do this worked, the pump to which I attached it didn't; in fact we have only recently managed to obtain a pump which does this operation in a satisfactory manner.

In 1969 and 1970 we carried on using the air pump as an extra scavenge pump, and the car manufacturers gradually sorted out their tanks. We ran the new separation system on the test bed towards the end of 1970 and it was fitted on Brabham's engine in Mexico, where it worked very well, though he was forced to retire while lying third when his normal pressure pump seized. It has a larger capacity than the previous scavenge pumps; it takes oil and air out of the engine in bucketfuls, puts them through a centrifuge and delivers the air out of one pipe and the oil out of another. It works from 4,000 engine r.p.m. upwards – below that the oil and air come back normally – so this should make it possible for the manufacturers to simplify their oil systems.

The big problem is that oil always manages to carry a fair amount of air around with it, and this is responsible for making the output of the pressure pump much lower than theoretically it should be. There is certainly something over 5 per cent of air in the oil, and it could be as much as 12 per cent. There are gears moving, there are camshafts whizzing around and there is the crankshaft, and so to get the blow-by out of the engine without taking half the oil with it is a major problem. As cornering forces have increased it has become increasingly difficult for oil to drain from the top to the bottom of an engine by gravity alone, because the oil level in tanks is now at least 50 degrees to the horizontal. Under those conditions on a continuous right-hander the level of the pump at the bottom of the sump is effectively above the bottom corner of the head. If it was only a matter of lubricating the bottom end it would be worthwhile considering using roller bearings, but you also need a fair amount of oil around the camshaft area, not only to lubricate the camshafts and the tappets but to cool the valve springs.

The biggest single problem we have had has been timing gears. We hope that we have stopped them breaking with our mechanism, but this does not mean that we have solved the problem; we have merely cushioned the effects of a shock loading which is far higher than we ever thought it would be.

The only way to be sure of getting over the timing gear problem would be to throw the gears away and use a cogged belt. I think cogged belt drive is fascinating. I think we would probably have a go at a cogged belt drive were it not for the fact that we would no longer be able to mount the engine in the way it is mounted at present. The size of the pulleys we would need on the camshafts would mean that the whole structure, and the idea of taking the loads through the rocker covers, would be wrecked.

The cams we were using in 1967 and 1969 had a rather ferocious stub torque. In order to save the timing gears we redesigned the camshafts to reduce the maximum torque which they required from around 36 lbs ft to 26 or thereabouts. In addition we were in a chronic valve spring situation at 9,500 r.p.m., and the engine wouldn't run above this speed on the original cam. The new cam improved this situation and we hoped that it would also help the gear train to survive, but subsequently we had a series of camshaft failures, which generally consisted of the exhaust camshaft on the left bank breaking near its rear end. This was not the place where you would have expected a camshaft to fail, and it was probably due to some obscure torsional problem. We tried reducing the hole in the middle of the camshaft and we gradually improved the situation – in fact several cases of alleged camshaft failure were due to people running out of oil, as a result of which pistons hit valves and broke sections out of camshafts.

In 1968 we produced some long intake trumpets for use on slow circuits, and these improved bottom end power output considerably. But apart from this the four-into-one exhaust system and the redesigned camshaft were the only major attempts we made to increase power output between 1967 and 1971, and the new camshafts didn't give any more power at 9,500 r.p.m., they merely allowed the engine to run faster. We didn't feel any real need to look for more power until we were beaten fair-and-square by the opposition, and this did not happen until 1970. Even then, we thought it was probably not a bad thing to lose the odd race. We didn't care very much for BRM winning at Spa because it meant we were beaten by a British team, but we didn't feel too badly about being beaten by Ferrari. Also, it was probably a good thing for motor racing that BRM did win at Spa; if they hadn't won a race in 1970 they might have packed it in. As Ferrari is one of Fiat's prestige outlets they are not likely to give up; they will go on putting more money into it until they are successful. There is, of course, some talk that Porsche will return to Formula 1, but I don't think they are too keen. They have had a look at grand prix racing before, and they have found the more sheltered pastures of sports car racing more productive.

During 1969 we had a number of valve spring failures. These were due to a marked decline in the quality of the springs, which more than offset the improvement we had made by changing our cam form. However, the real problem was the same extraordinary load which caused our valve gear trouble. It was probably something very similar which caused the four-valve Repco engine to be withdrawn from racing; despite frequent changes of material, number six inlet tappet failed repeatedly. Coventry Climax were unable to drive their four-valve engine with chains and had to convert to gears – and they had a fair amount of trouble with them. I think the combination of four valves, increased camshaft overlap, high instantaneous torques and V8 firing intervals probably gives rise to a rather nasty torsional condition. We have recorded instantaneous torques of over 300 lbs ft, whereas the theoretical maximum is 26 lbs ft at 10,000 r.p.m. All we have done, in strengthening and supporting the gears, is to transfer the failure from one component to another. However, we hope that we have now overcome the problem, even if we haven't solved it, by designing a mechanism which will absorb these shock loads. It is very simple really – it consists of a pair of gears linked by twelve tiny quill shafts – but it does seem to have reduced the loads to something near the theoretical maximum. The biggest problem, of course, was fitting it into the available space. By the end of 1970 we had fitted this mechanism in seven engines and none of them had had any major failures, so we are hoping that our valve gear problems are a thing of the past.

From the outset we designed the nose of the crankshaft with a view to fitting a damper. In fact we had relatively few crankshaft failures until 1970, and those were due to a manufacturing problem. But in 1970 a torsional investigation was carried out and it was found that we had a large amplitude torsional oscillation at about 6,000 r.p.m. – as a result of which we designed a damper, to be made available for the 1971 engines. The main reason for doing this was to increase our chances of running faster.

The history of our 1970 crankshaft failures is very interesting. On all our engines, starting with the SCA, there has been a tendency for particles of nitrided metal to come out of the cheeks of the crank-pins and embed themselves in the relatively soft surface of the con-rod. This has acted as a superb nitride-hardened slab and has cut away the cheeks of the crank-pins by as much as 60 thou. In an attempt to cure this we tried to lap off about a thou and a half a side from the cheeks of the crank-pins. The shafts are nitrided when they are made, leaving four thou on the diameter of the mains; the crank-pins are made to about bottom limit.

After nitriding the main bearing journals are all ground, but the crank-pins are just polished with emery to remove the friable top layer. We managed to grind two thou off each side on certain cranks, and as a result we asked the camshaft manufacturers to do this as a production item. We told them that they didn't need to grind the pins; all we wanted them to do was to grind two thou off each cheek. On a big crankshaft grinder this would require very great skill, and the manufacturers decided that they would leave the crank-pins four thou oversize and form a wheel with the width to take two thou off each side. This seemed to work quite well, except that it was difficult to position the wheel sufficiently accurately to take two thou off each side – and therefore there was a tendency for the odd crank to have four thou off one side and nothing off the other. In addition it would appear that the journals were ground before nitriding with a wheel which had too large a radius on the corners. After nitriding they came along with a wheel which had the correct radius on the corners, with the result that the nitrided layer was still at full thickness along the journal with approximately two thou along the flanks; because of the differences in radius it went through the nitrided layer into the core in the middle of the radius, especially on those cranks where there had been some eccentricity and wheel off-set. This dropped the life of the crank-shaft from something in the region of 100 hours down to about three hours; I think 19 hours was the longest that any of these cranks lasted. We carried out a scratch test and found that on a lot of the cranks we couldn't mark the journal or the cheek side with an ordinary scriber, but we were able to scratch the middle of the radius quite easily. As a result we were forced to withdraw the whole batch of crankshafts. Fortunately we had another batch of what were nominally spare cranks coming along, and we managed to get these ground to the finished size and nitrided; we lapped the mains and the pins ourselves this time.

Some time ago we decided that the pistons were a bit marginal, and we strengthened them up in various ways about the middle of 1969. I made some detail changes to the piston design on both the Formula 2 and Formula 1 and these worked very well on the FVA but not so well on the DFV. We then altered our land clearance because we thought it was this which was causing the pistons to seize up. In practice we think that our oil consumption problems and a certain amount of power loss were due to the pistons rattling around in the bore. We made a further series of modifications to the pistons, but we really needed more time for testing. The problem was to have enough pistons to replace those that were in danger of breaking or had already scuffed up. It was very difficult to know whether to go ahead with the old pistons, which we knew would work reasonably, or to make a lot of the new type, which could be better and could be worse.

The state of the art of engine design is still very rudimentary. It is quite possible that in the near future somebody will produce an engine which gives a lot more power than anything which is running at present. It all depends on someone having a good idea and executing it well in detail. There is always a chance that Ferrari will make a big break-through. They are doing a lot of instrumenting of engines, and provided they don't become too confused by the scientific approach this could be of great assistance to them. It also helps that they have to make only two or three engines for any given race, and therefore can afford to experiment with different configurations. There is a fair amount of trial and error in anything of this sort, and he who tries the greatest number of alternatives has the best chance of finding out the right combination.

For instance, by going unconventional, as Ferrari have on their flat-12 engine, using a combination of roller bearings and plain bearings, there is a strong chance of gaining a few horsepower. However, it could take a lot of development to make something of this sort reliable, although Ferrari are in a better position to carry out the research and development than we are; as we are selling engines to people we have an obligation to try to make them reliable and not phenomenally expensive.

Over the years there has been a frantic increase in the extent to which the engines are caned. In 1967 a fuel consumption of 8·8 m.p.g. was fairly common, but in 1970 we were down to 6 m.p.g. with the same metering cam. This is a reduction of over 25 per cent, and shows that the engines are being held at high revs and full throttle for a far greater proportion of the time. Part of this is due to chassis and tyre improvements, and partly due to the fact that everybody is now that much more competitive. In 1967 it was possible to win at a canter.

The reliability of any given engine depends largely on how hard it is caned, and there are difficulties in deciding how long an engine should last. How many short-life components should we use in our engines? Ferrari may well decide that a piston life of 400 kilometres is adequate (this would mean changing pistons after every race). If he does then we will virtually be obliged to do the same, but the fact that the engine has to be stripped and rebuilt frequently, simply in order to change pistons, will naturally add to the cost. The more highly developed and competitive a formula, the higher is liable to be the cost of maintaining engines for it.

When we designed the DFV we did not give much consideration to the number of years for which it might remain competitive. We merely designed an engine which we felt would be capable of winning some Formula 1 races. Nor did we give any thought to the possibility of producing the engine in quantity. We merely planned to make five engines for one team and see how we got on. However, the terms of our original agreement weren't particularly rewarding to us in the long term, and so when the engine had shown that it was capable of winning and that it wasn't too unreliable, we felt that if we could sell a few more it would be of financial advantage to us. We also felt that in doing this we would be providing something of a service for motor racing, because there was a chronic shortage of cars at that time due to a shortage of competitive engines.

Throughout the first five years of the DFV's life we carried out very little development work with a view to getting more power. Although we had worked quite hard to try and make it more reliable – there was no point in attempting to get more power until we had done this – it was not until January 1971 that we began to look for ways of increasing the output.

The DFV engine had given a lower specific output than the FVA, and we decided to try to find out why. It could, we felt, be something to do with the difference in valve angles – 32 degrees on the DFV as against 40 on the FVA – or it could be that the DFV cylinder head had never been made as it was designed. We had only discovered this a short while previously when we sectioned a damaged head. There were supposed to be water passages on either side of the sparking plug, but in fact there was solid metal all round it. Either the passages had been too small to cast, or the foundry people had chosen to ignore them. We also decided to investigate port design; having always felt that the ports were a little on the small side, we decided to try a slightly different shape.'

This and other development work on the DFV was carried out under Keith Duckworth's direction during the early months of 1971 in answer to the growing challenge being offered by the three rival 12-cylinder engines.

It was known that Ferrari were preparing a new and potentially better chassis to carry their latest roller-bearing flat-12 engine; that Matra-Simca were so keen to find more power from their V12 that they had concentrated most of their between-seasons development time specifically on this task rather than on major chassis improvements; and that BRM were preparing a new shorter-stroke version of their V12 engine, aimed at giving their well-handling car considerably more power through the use of higher engine revolutions.

There was no doubt, therefore, that the power battle would be closer-fought – and probably more exciting to watch – than at any time since the inception of the 3 litre Formula 1 in 1966. That the formula had proved a resounding success there was no question, for grand prix racing had never been so international. However, whether Formula 1 could have survived, far less thrived, without the existence of the DFV engine is very doubtful. This is the measure of the contribution made by Cosworth Engineering and the Ford Motor Company to the world of motor racing. They shook hands on a deal worth £100,000 but the true value of that handshake to motorsport has been incalculable, and will remain so for as long as the Sweet Thunder of the DFV echoes around the racing circuits of the world.

On First Looking Into Chapman's Lotus

Much have I travell'd in the realms of gold,
  And many goodly crowns and pinions seen;
  Round many hairy circuits have I been
Which bards in fealty to racing hold.
Oft of one great equipe had I been told
  That deep-brow'd Chapman ruled as his demesne:
  Yet never did I hear its gearbox scream
Till Zandvoort, where it spake out loud and bold:
Then felt I like some watcher of the skies
  When a new planet swims into his ken;
Or like stout Henry, when with eagle eyes
  He stared at his Tin Lizzie – and all his men
Look'd at each other with a wild surmise –
  Silent, upon a peak in Michigan.

This tongue-in-cheek paraphrase of Keats' *On First Looking into Chapman's Homer* was written by a Ford Public Relations man on the morning after the historic win at Zandvoort.

Appendix I

Index of Ford DFV and DFW Engines

Engine number	Delivery date	Customer	Subsequent history
701	—	Development engine	Written off during initial testing
702	28/4/1967	Team Lotus	Written off by Baghetti, Italian GP, September 1967
703	26/5/1967	Team Lotus	Converted to DFW 703 with 2½ litre Tasman connecting rods
704	2/6/1967	Team Lotus	Converted to DFW 704 with 2½ litre Tasman rods.
705	23/6/1967	Team Lotus	Converted to DFW 705 with 2½ litre Tasman rods. Later reconverted to DFV and sold to John Love
706	28/7/1967	Team Lotus	Became Ford Motor Company's exhibition engine
707	12/10/1967	Team Lotus	Sold to Rob Walker, destroyed in garage fire, March 1968
801	24/11/1967	Team Lotus	Updated and redesignated DFV 901, August 1969
802	6/12/1967	Tyrrell Racing Organisation	So to Silvio Moser, July 1969. Updated and redesignated DFV 902, April 1970
803	22/12/1967	Team Lotus	Converted to 2½ litre Tasman DFW 803, December 1968
804	23/1/1968	Alan Mann Racing	Sold to J. W. Automotive, January 1970. Sold to Tom Wheatcroft, March 1970
805	1/3/1968	McLaren Racing	Updated and redesignated DFV 905, August 1969

Engine number	Delivery date	Customer	Subsequent history
806	13/3/1968	McLaren Racing	Updated and redesignated DFV 906, December 1969
807	14/3/1968	Tyrrell Racing Organisation	Updated and redesignated DFV 907, March 1970
808	3/4/1968	Alan Mann Racing	Updated and redesignated DFV 908, September 1969. Sold to Frank Williams, October 1969. Sold to Performance Equipment Co., South Africa, for John Love, 1970
809	11/4/1968	Rob Walker	Updated and redesignated DFV 909, August 1969
810	25/4/1968	McLaren Racing	Sold to van Rooyen 1969, modified by van Rooyen. Reconverted to DFV 910 by Cosworth, August 1970 and sold to Guy Ligier for Le Mans car
811	26/4/1968	Team Lotus	Updated and redesignated DFV 911, August 1969
812	1/5/1968	Tyrrell Racing Organisation	Updated and redesignated DFV 912, January 1970. Sold Jo Bonnier
813	4/5/1968	McLaren Racing	Sold to van Rooyen 1969, modified by van Rooyen. Reconverted to DFV 913 by Cosworth, August 1970. Sold to Tyrrell Racing Organisation
814	8/5/1968	Team Lotus	Sold to Pete Lovely, May 1969. Still designated DFV 814 but has 'Series 9' modifications
815	10/5/1968	Alan Mann Racing	Sold to J. W. Automotive, 1969.Sold to STP, updated and redesignated DFV 915, April 1970, and sold to Antique Automobiles, May 1970
816	16/5/1968	Alan Mann Racing	Sold to Alan de Cadanet for McLaren sports car, updated and redesignated DFV 916, February 1970
817	18/5/1968	Tyrrell Racing Organisation	Sold to Antique Automobiles, June 1969. Sold to March Engineering, updated and redesignated DFV 917, February 1970. Sold to STP

Engine number	Delivery date	Customer	Subsequent history
818	29/5/1968	Team Lotus	Sold to Dave Charlton, 1970. Still designated DFV 818 but has 'Series 9' modifications
819	30/5/1968	McLaren Racing	Updated and redesignated DFV 919, August 1969
820	15/6/1968	Rob Walker	Updated and redesignated DFV 920, September, 1969
821	5/8/1968	Team Lotus	Originally built as 2½ litre DFW. Converted and updated to DFV 921, August 1969
822	3/9/1968	Mitsui Trading Co.	Not seen since!
823	20/9/1968	Mitsui Trading Co.	Not seen since!
824	16/10/1968	Frank Williams	Originally built as 2½ litre DFW. Converted to DFV, March 1969. Updated and redesignated DFV 924, September 1969
825	5/11/1968	Frank Williams	Originally built as 2½ litre DFW. Converted to DFV, March 1969. Updated and redesignated DFV 925, September 1969
826	6/1/1969	Motor Racing Developments	Updated and redesignated DFV 926, September 1969
927	21/2/1969	Tyrrell Racing Organisation	Sold to March Engineering, May 1970
928	4/3/1969	McLaren Racing	
929	10/3/1969	Team Lotus	
930	21/2/1969	Motor Racing Developments	Sold to Gus Hutchinson for Formula A, 1970. Sold to Team Surtees, 1971
931	15/3/1969	J. W. Automotive	
932	25/3/1969	Rob Walker	
933	26/3/1969	McLaren Racing	
934	26/3/1969	Team Lotus	
935	11/4/1969	Tyrrell Racing Organisation	Sold to March Engineering, February 1970
936	25/4/1969	McLaren Racing	Sold to Team Surtees, December 1969
937	25/4/1969	Motor Racing Developments	
938	1/5/1969	Tyrrell Racing Organisation	Sold to Jo Bonnier, 1970

Engine number	Delivery date	Customer	Subsequent history
939	30/4/1969	Motor Racing Developments	
940	16/5/1969	Tyrrell Racing Organisation	
941	30/6/1969	J. W. Automotive	
942	11/6/1969	Tyrrell Racing Organisation	Sold to March Engineering, May 1970
943	2/7/1969	Kurt Ahrens (Mercedes-Benz)	
944	14/7/1969	Performance Equipment Co. (John Love)	
945	12/7/1969	Team Lotus	
946	25/8/1969	Tyrrell Racing Organisation	
947	29/9/1969	Newspaper 'La Razon' (Argentina)	Bought for Argentinian Berta sports car
948	4/11/1969	Ford of Germany (MRD – Stommelen)	
949	12/11/1969	Ford of Germany (MRD – Stommelen)	
950	19/2/1970	Team Surtees	Updated and redesignated DFV 050, January 1971
051	15/7/1969	Development engine	
052	3/3/1970	Team Lotus	
053	2/3/1970	Tyrrell Racing Organisation	
054	5/3/1970	Tyrrell Racing Organisation	
055	24/3/1970	Champlin Racing (Gus Hutchinson)	
056	16/3/1970	Motor Racing Developments	
057	20/3/1970	Tyrrell Racing Organisation	
058	17/3/1970	Team Lotus	
059	24/3/1970	Team Lotus	
060	17/3/1970	McLaren Racing	
061	11/3/1970	McLaren Racing	
062	13/3/1970	Team Surtees	
063	25/4/1970	Tyrrell Racing Organisation	
064	24/3/1970	March Engineering	

Engine number	Delivery date	Customer	Subsequent history
065	5/5/1970	March Engineering	
066	2/3/1971	Tyrrell Racing Organisation	
067	24/4/1970	Frank Williams	
068	2/4/1971	Team Surtees	
069	8/7/1970	Umberto Maglioli	Bought for Panther sports car
070	15/4/1971	Motor Racing Developments	
071	9/5/1970	STP Corporation	
072	27/3/1971	McLaren Racing	
073	18/6/1970	STP Corporation	
074	16/4/1971	Team Lotus	
075	16/4/1971	Rob Walker	
076	30/4/1971	Frank Williams	
077	15/4/1971	Tyrell Racing Organisation	
078	7/4/71	Team Surtees	
079	30/4/1971	Motor Racing Developments	
080	5/5/1971	McLaren Racing	
082	5/5/1971	Team Lotus	
083	5/5/1971	March Engineering	

NOTE: Serial numbers 703DEV, 081 and 091 are allocated to development engines other than those listed above.

NOTE: Engines with even serial numbers from 066 onwards were allocated to customers in 1970 but due to a production cutback were not manufactured until 1971. These engines, therefore, were built to 1971 specification. Serial numbers from 1970 onwards are strictly numerical and do not necessarily relate to specification.

Appendix 2

World championship scoresheet, 1966–70

1966

MONACO GRAND PRIX – Monte Carlo			Points
1	Jackie Stewart	BRM 2 litre V8	9
2	Lorenzo Bandini	Ferrari 2·4 litre V6	6
3	Graham Hill	BRM 2 litre V8	4
4	Bob Bondurant	BRM 2 litre V8	3
	No other finishers		

BELGIAN GRAND PRIX – Spa-Francorchamps			
1	John Surtees	Ferrari V12	9
2	Jochen Rindt	Cooper-Maserati V12	6
3	Lorenzo Bandini	Ferrari 2·4 litre V6	4
4	Jack Brabham	Repco-Brabham V8	3
5	Richie Ginther	Cooper-Maserati V12	2
	No other finishers		

FRENCH GRAND PRIX – Reims			
1	Jack Brabham	Repco Brabham V8	9
2	Mike Parkes	Ferrari V12	6
3	Denny Hulme	Repco Brabham V8	4
4	Jochen Rindt	Cooper-Maserati V12	3
5	Dan Gurney	Eagle-Climax 2·5 litre 4	2
6	John Taylor	Brabham-BRM 2 litre V8	1

BRITISH GRAND PRIX – Brands Hatch			
1	Jack Brabham	Repco Brabham V8	9
2	Denny Hulme	Repco Brabham V8	6
3	Graham Hill	BRM 2 litre V8	4
4	Jim Clark	Lotus-Climax 2 litre V8	3
5	Jochen Rindt	Cooper-Maserati V12	2
6	Bruce McLaren	McLaren-Serenissima V8	1

DUTCH GRAND PRIX – Zandvoort

			Points
1	Jack Brabham	Repco Brabham V8	9
2	Graham Hill	BRM 2 litre V8	6
3	Jim Clark	Lotus-Climax 2 litre V8	4
4	Jackie Stewart	BRM 2 litre V8	3
5	Mike Spence	Lotus-BRM 2 litre V8	2
6	Lorenzo Bandini	Ferrari V12	1

GERMAN GRAND PRIX – Nurburgring

1	Jack Brabham	Repco Brabham V8	9
2	John Surtees	Cooper-Maserati V12	6
3	Jochen Rindt	Cooper Maserati V12	4
4	Graham Hill	BRM 2 litre V8	3
5	Jackie Stewart	BRM 2 litre V8	2
6	Lorenzo Bandini	Ferrari V12	1

ITALIAN GRAND PRIX – Monza

1	Lodovico Scarfiotti	Ferrari V12	9
2	Mike Parkes	Ferrari V12	6
3	Denny Hulme	Repco Brabham V8	4
4	Jochen Rindt	Cooper-Maserati V12	3
5	Mike Spence	Lotus-BRM 2 litre V8	2
6	Bob Anderson	Brabham-Climax 2·7 litre 4	1

UNITED STATES GRAND PRIX – Watkins Glen

1	Jim Clark	BRM H16	9
2	Jochen Rindt	Cooper-Maserati V12	6
3	John Surtees	Cooper-Maserati V12	4
4	Jo Siffert	Cooper-Maserati V12	3
5	Bruce McLaren	McLaren-Ford V8	2
6	Peter Arundell	Lotus-Climax 2 litre V8	1

MEXICAN GRAND PRIX – Mexico City

1	John Surtees	Cooper-Maserati V12	9
2	Jack Brabham	Repco Brabham V8	6
3	Denny Hulme	Repco Brabham V8	4
4	Richie Ginther	Honda V12	3
5	Dan Gurney	Eagle-Climax 2·7 litre 4	2
6	Jo Bonnier	Cooper-Maserati V12	1

1966 World Champion driver Jack Brabham
1966 World Champion car Repco Brabham V8

1967

SOUTH AFRICAN GRAND PRIX – Kyalami

1	Pedro Rodriguez	Cooper-Maserati V12	9
2	John Love	Cooper-Climax 2·7 litre 4	6
3	John Surtees	Honda V12	4
4	Denny Hulme	Repco Brabham V8	3
5	Bob Anderson	Brabham-Climax 2·7 litre 4	2
6	Jack Brabham	Repco Brabham V8	1

MONACO GRAND PRIX – Monte Carlo

			Points
1	Denny Hulme	Repco Brabham V8	9
2	Graham Hill	Lotus-BRM 2 litre V8	6
3	Chris Amon	Ferrari V12	4
4	Bruce McLaren	McLaren-BRM 2 litre V8	3
5	Pedro Rodriguez	Cooper-Maserati V12	2
6	Mike Spence	BRM H16	1

DUTCH GRAND PRIX – Zandvoort

1	**Jim Clark**	**Lotus-Ford V8 DFV 703**	**9**
2	Jack Brabham	Repco Brabham V8	6
3	Denny Hulme	Repco Brabham V8	4
4	Chris Amon	Ferrari V12	3
5	Mike Parkes	Ferrari V12	2
6	Lodovico Scarfiotti	Ferrari V12	1

BELGIAN GRAND PRIX – Spa-Francorchamps

1	Dan Gurney	Eagle-Weslake V12	9
2	Jackie Stewart	BRM H16	6
3	Chris Amon	Ferrari V12	4
4	Jochen Rindt	Cooper-Maserati V12	3
5	Mike Spence	BRM H16	2
6	**Jim Clark**	**Lotus-Ford V8 DFV 703**	**1**

FRENCH GRAND PRIX – Le Mans

1	Jack Brabham	Repco Brabham V8	9
2	Denny Hulme	Repco Brabham V8	6
3	Jackie Stewart	BRM 2·1 litre V8	4
4	Jo Siffert	Cooper-Maserati V12	3
5	Chris Irwin	BRM H16	2
6	Pedro Rodriguez	Cooper-Maserati V12	1

BRITISH GRAND PRIX – Silverstone

1	**Jim Clark**	**Lotus-Ford V8 DFV 703**	**9**
2	Denny Hulme	Repco Brabham V8	6
3	Chris Amon	Ferrari V12	4
4	Jack Brabham	Repco Brabham V8	3
5	Pedro Rodriguez	Cooper-Maserati V12	2
6	John Surtees	Honda V12	1

GERMAN GRAND PRIX – Nurburgring

1	Denny Hulme	Repco Brabham V8	9
2	Jack Brabham	Repco Brabham V8	6
3	Chris Amon	Ferrari V12	4
4	John Surtees	Honda V12	3
5	Jo Bonnier	Cooper-Maserati V12	2
6	Guy Ligier	Cooper-Maserati V12	1

CANADIAN GRAND PRIX – Mosport			Points
1	Jack Brabham	Repco Brabham V8	9
2	Denny Hulme	Repco Brabham V8	6
3	Dan Gurney	Eagle-Weslake V12	4
4	**Graham Hill**	**Lotus-Ford V8 DFV 704**	**3**
5	Mike Spence	BRM H16	2
6	Chris Amon	Ferrari V12	1

ITALIAN GRAND PRIX – Monza			
1	John Surtees	Honda V12	9
2	Jack Brabham	Repco Brabham V8	6
3	**Jim Clark**	**Lotus-Ford V8 DFV 704**	**4**
4	Jochen Rindt	Cooper-Maserati V12	3
5	Mike Spence	BRM H16	2
6	Jacky Ickx	Cooper-Maserati V12	1

UNITED STATES GRAND PRIX – Watkins Glen			
1	**Jim Clark**	**Lotus-Ford V8 DFV 707**	**9**
2	**Graham Hill**	**Lotus-Ford V8 DFV 705**	**6**
3	Denny Hulme	Repco Brabham V8	4
4	Jo Siffert	Cooper-Maserati V12	3
5	Jack Brabham	Repco Brabham V8	2
6	Jo Bonnier	Cooper-Maserati V12	1

MEXICAN GRAND PRIX – Mexico City			
1	**Jim Clark**	**Lotus-Ford V8 DFV 704**	**9**
2	Jack Brabham	Repco Brabham V8	6
3	Denny Hulme	Repco Brabham V8	4
4	John Surtees	Honda V12	3
5	Mike Spence	BRM H16	2
6	Pedro Rodriguez	Cooper-Maserati V12	1

1967 World Champion driver	Denny Hulme
1967 World Champion car	Repco Brabham V8

1968

SOUTH AFRICAN GRAND PRIX – Kyalami			
1	**Jim Clark**	**Lotus-Ford V8 DFV 803**	**9**
2	**Graham Hill**	**Lotus-Ford V8 DFV 801**	**6**
3	Jochen Rindt	Repco Brabham V8	4
4	Chris Amon	Ferrari V12	3
5	Denny Hulme	McLaren-BRM V12	2
6	Jean-Pierre Beltoise	Matra-Ford 1·6 litre 4	1

SPANISH GRAND PRIX – Jarama			
1	**Graham Hill**	**Lotus-Ford V8 DFV 704**	**9**
2	**Denny Hulme**	**McLaren-Ford V8 DFV 810**	**6**
3	Brian Redman	Cooper-BRM V12	4
4	Lodovico Scarfiotti	Cooper-BRM V12	3
5	**Jean-Pierre Beltoise**	**Matra-Ford V8 DFV 802**	**2**
	No other finishers		

MONACO GRAND PRIX – Monte Carlo

			Points
1	**Graham Hill**	**Lotus-Ford V8 DFV 814**	**9**
2	Richard Attwood	BRM V12	6
3	Lucien Bianchi	Cooper-BRM V12	4
4	Lodovico Scarfiotti	Cooper-BRM V12	3
5	**Denny Hulme**	**McLaren-Ford V8 DFV 810**	**2**
	No other finishers		

BELGIAN GRAND PRIX – Spa-Francorchamps

1	**Bruce McLaren**	**McLaren-Ford V8 DFV 805**	**9**
2	Pedro Rodriguez	BRM V12	6
3	Jacky Ickx	Ferrari V12	4
4	**Jackie Stewart**	**Matra-Ford V8 DFV 807**	**3**
5	**Jackie Oliver**	**Lotus-Ford V8 DFV 803**	**2**
6	Lucien Bianchi	Cooper-BRM V12	1

DUTCH GRAND PRIX – Zandvoort

1	**Jackie Stewart**	**Matra-Ford V8 DFV 812**	**9**
2	Jean-Pierre Beltoise	Matra V12	6
3	Pedro Rodriguez	BRM V12	4
4	Jacky Ickx	Ferrari V12	3
5	Silvio Moser	Repco Brabham V8	2
6	Chris Amon	Ferrari V12	1

FRENCH GRAND PRIX – Rouen

1	Jacky Ickx	Ferrari V12	9
2	John Surtees	Honda V12	6
3	**Jackie Stewart**	**Matra-Ford V8 DFV 812**	**4**
4	Vic Elford	Cooper-BRM V12	3
5	**Denny Hulme**	**McLaren-Ford V8 DFV 805**	**2**
6	Piers Courage	BRM V12	1

BRITISH GRAND PRIX – Brands Hatch

1	**Jo Siffert**	**Lotus-Ford V8 DFV 809**	**9**
2	Chris Amon	Ferrari V12	6
3	Jacky Ickx	Ferrari V12	4
4	**Denny Hulme**	**McLaren Ford V8 DFV 819**	**3**
5	John Surtees	Honda V12	2
6	**Jackie Stewart**	**Matra-Ford V8 DFV 817**	**1**

GERMAN GRAND PRIX – Nurburgring

1	**Jackie Stewart**	**Matra-Ford V8 DFV 812**	**9**
2	**Graham Hill**	**Lotus-Ford V8 DFV 811**	**6**
3	Jochen Rindt	Repco Brabham V8	4
4	Jacky Ickx	Ferrari V12	3
5	Jack Brabham	Repco Brabham V8	2
6	Pedro Rodriguez	BRM V12	1

ITALIAN GRAND PRIX – Monza			Points
1	**Denny Hulme**	**McLaren-Ford V8 DFV 819**	**9**
2	**Johnny Servoz-Gavin**	**Matra-Ford V8 DFV 817**	**6**
3	Jacky Ickx	Ferrari V12	4
4	Piers Courage	BRM V12	3
5	Jean-Pierre Beltoise	Matra V12	2
6	Jo Bonnier	McLaren-BRM V12	1

CANADIAN GRAND PRIX – St. Jovite			
1	**Denny Hulme**	**McLaren-Ford V8 DFV 819**	**9**
2	**Bruce McLaren**	**McLaren-Ford V8 DFV 805**	**6**
3	Pedro Rodriguez	BRM V12	4
4	**Graham Hill**	**Lotus-Ford V8 DFV 801**	**3**
5	Vic Elford	Cooper-BRM V12	2
6	**Jackie Stewart**	**Matra-Ford V8 DFV 807**	**1**

UNITED STATES GRAND PRIX – Watkins Glen			
1	**Jackie Stewart**	**Matra-Ford V8 DFV 817**	**9**
2	**Graham Hill**	**Lotus-Ford V8 DFV 804**	**6**
3	John Surtees	Honda V12	4
4	**Dan Gurney**	**McLaren-Ford V8 DFV 813**	**3**
5	**Jo Siffert**	**Lotus-Ford V8 DFV 809**	**2**
6	**Bruce McLaren**	**McLaren-Ford V8 DFV 805**	**1**

MEXICAN GRAND PRIX – Mexico City			
1	**Graham Hill**	**Lotus-Ford V8 DFV 803**	**9**
2	**Bruce McLaren**	**McLaren-Ford V8 DFV 810**	**6**
3	**Jackie Oliver**	**Lotus-Ford V8 DFV 814**	**4**
4	Pedro Rodriguez	BRM V12	3
5	Jo Bonnier	Honda V12	2
6	**Jo Siffert**	**Lotus-Ford V8 DFV 820**	**1**

1968 World Champion driver	**Graham Hill**
1968 World Champion car	**Lotus-Ford V8 DFV**

1969

SOUTH AFRICAN GRAND PRIX – Kyalami			
1	**Jackie Stewart**	**Matra-Ford V8 DFV 817**	**9**
2	**Graham Hill**	**Lotus-Ford V8 DFV 801**	**6**
3	**Denny Hulme**	**McLaren-Ford V8 DFV 805**	**4**
4	**Jo Siffert**	**Lotus-Ford V8 DFV 809**	**3**
5	**Bruce McLaren**	**McLaren-Ford V8 DFV 806**	**2**
6	**Jean-Pierre Beltoise**	**Matra-Ford V8 DFV 802**	**1**

SPANISH GRAND PRIX – Montjuich			
1	**Jackie Stewart**	**Matra-Ford V8 DFV 802**	**9**
2	**Bruce McLaren**	**McLaren-Ford V8 DFV 936**	**6**
3	**Jean-Pierre Beltoise**	**Matra-Ford V8 DFV 817**	**4**
4	**Denny Hulme**	**McLaren-Ford V8 DFV 928**	**3**
5	John Surtees	BRM V12	2
6	**Jacky Ickx**	**Brabham-Ford V8 DFV 937**	**1**

MONACO GRAND PRIX – Monte Carlo

			Points
1	Graham Hill	Lotus-Ford V8 DFV 811	9
2	Piers Courage	Brabham-Ford V8 DFV 825	6
3	Jo Siffert	Lotus-Ford V8 DFV 809	4
4	Richard Attwood	Lotus-Ford V8 DFV 818	3
5	Bruce McLaren	McLaren-Ford V8 DFV 936	2
6	Denny Hulme	McLaren-Ford V8 DFV 928	1

DUTCH GRAND PRIX – Zandvoort

1	Jackie Stewart	Matra-Ford V8 DFV 940	9
2	Jo Siffert	Lotus-Ford V8 DFV 932	6
3	Chris Amon	Ferrari V12	4
4	Denny Hulme	McLaren-Ford V8 DFV 929	3
5	Jacky Ickx	Brabham-Ford V8 DFV 939	2
6	Jack Brabham	Brabham-Ford V8 DFV 826	1

FRENCH GRAND PRIX – Clermont-Ferrand

1	Jackie Stewart	Matra-Ford V8 DFV 938	9
2	Jean-Pierre Beltoise	Matra-Ford V8 DFV 927	6
3	Jacky Ickx	Brabham-Ford V8 DFV 937	4
4	Bruce McLaren	McLaren-Ford V8 DFV 936	3
5	Vic Elford	McLaren-Ford V8 DFV 817	2
6	Graham Hill	Lotus-Ford V8 DFV 934	1

BRITISH GRAND PRIX – Silverstone

1	Jackie Stewart	Matra-Ford V8 DFV 812	9
2	Jacky Ickx	Brabham-Ford V8 DFV 930	6
3	Bruce McLaren	McLaren-Ford V8 DFV 933	4
4	Jochen Rindt	Lotus-Ford V8 DFV 945	3
5	Piers Courage	Brabham-Ford V8 DFV 824	2
6	Vic Elford	McLaren-Ford V8 DFV 817	1

GERMAN GRAND PRIX – Nurburgring

1	Jacky Ickx	Brabham-Ford V8 DFV 939	9
2	Jackie Stewart	Matra-Ford V8 DFV 935	6
3	Bruce McLaren	McLaren-Ford V8 DFV 933	4
4	Graham Hill	Lotus-Ford DFV 934	3
5	Jo Siffert	Lotus-Ford V8 DFV 932	2
6	Jean-Pierre Beltoise	Matra-Ford V8 DFV 940	1

ITALIAN GRAND PRIX – Monza

1	Jackie Stewart	Matra-Ford V8 DFV 946	9
2	Jochen Rindt	Lotus-Ford V8 DFV 921	6
3	Jean-Pierre Beltoise	Matra-Ford V8 DFV 927	4
4	Bruce McLaren	McLaren-Ford V8 DFV 936	3
5	Piers Courage	Brabham-Ford V8 DFV 824	2
6	Pedro Rodriguez	Ferrari V12	1

CANADIAN GRAND PRIX – Mosport			Points
1	**Jacky Ickx**	**Brabham-Ford V8 DFV 926**	**9**
2	**Jack Brabham**	**Brabham-Ford V8 DFV 937**	**6**
3	**Jochen Rindt**	**Lotus-Ford V8 DFV 921**	**4**
4	**Jean-Pierre Beltoise**	**Matra-Ford V8 DFV 927**	**3**
5	**Bruce McLaren**	**McLaren-Ford V8 DFV 933**	**2**
6	**Johnny Servoz-Gavin**	**Matra-Ford V8 DFV 942**	**1**

UNITED STATES GRAND PRIX Watkins Glen			
1	**Jochen Rindt**	**Lotus-Ford V8 DFV 921**	**9**
2	**Piers Courage**	**Brabham-Ford V8 DFV 924**	**6**
3	John Surtees	BRM V12	4
4	**Jack Brabham**	**Brabham-Ford V8 DFV 937**	**3**
5	Pedro Rodriguez	Ferrari V12	2
6	**Silvio Moser**	**Brabham-Ford V8 DFV 802**	**1**

MEXICAN GRAND PRIX – Mexico City			
1	**Denny Hulme**	**McLaren-Ford V8 DFV 933**	**9**
2	**Jacky Ickx**	**Brabham-Ford V8 DFV 939**	**6**
3	**Jack Brabham**	**Brabham-Ford V8 DFV 937**	**4**
4	**Jackie Stewart**	**Matra-Ford V8 DFV 946**	**3**
5	**Jean-Pierre Beltoise**	**Matra-Ford V8 DFV 935**	**2**
6	Jackie Oliver	BRM V12	1

1969 World Champion driver — **Jackie Stewart**
1969 World Champion car — **Matra-Ford V8 DFV**

1970

SOUTH AFRICAN GRAND PRIX – Kyalami			
1	**Jack Brabham**	**Brabham-Ford V8 DFV 948**	**9**
2	**Denny Hulme**	**McLaren-Ford V8 DFV 906**	**6**
3	**Jackie Stewart**	**March-Ford V8 DFV 053**	**4**
4	Jean-Pierre Beltoise	Matra-Simca V12	3
5	**John Miles**	**Lotus-Ford V8 DFV 901**	**2**
6	**Graham Hill**	**Lotus-Ford V8 DFV 920**	**1**

SPANISH GRAND PRIX – Jarama			
1	**Jackie Stewart**	**March-Ford V8 DFV 054**	**9**
2	**Bruce McLaren**	**McLaren-Ford V8 DFV 933**	**6**
3	**Mario Andretti**	**March-Ford V8 DFV 917**	**4**
4	**Graham Hill**	**Lotus-Ford V8 DFV 909**	**3**
5	**Johnny Servoz-Gavin**	**March-Ford V8 DFV 057**	**2**
	No other finishers		

MONACO GRAND PRIX – Monte Carlo			
1	**Jochen Rindt**	**Lotus-Ford V8 DFV 934**	**9**
2	**Jack Brabham**	**Brabham-Ford V8 DFV 937**	**6**
3	Henri Pescarolo	Matra-Simca V12	4
4	**Denny Hulme**	**McLaren-Ford V8 DFV 928**	**3**
5	**Graham Hill**	**Lotus-Ford V8 DFV 921**	**2**
6	Pedro Rodriguez	BRM V12	1

BELGIAN GRAND PRIX – Spa-Francorchamps			Points
1	Pedro Rodriguez	BRM V12	9
2	**Chris Amon**	**March-Ford V8 DFV 907**	6
3	Jean-Pierre Beltoise	Matra-Simca V12	4
4	Ignazio Giunti	Ferrari F12	3
5	**Rolf Stommelen**	**Brabham-Ford V8 DFV 949**	2
6	Henri Pescarolo	Matra-Simca V12	1

DUTCH GRAND PRIX – Zandvoort

1	**Jochen Rindt**	**Lotus-Ford V8 DFV 945**	9
2	**Jackie Stewart**	**March-Ford V8 DFV 057**	6
3	Jacky Ickx	Ferrari F12	4
4	Clay Regazzoni	Ferrari F12	3
5	Jean-Pierre Beltoise	Matra-Simca V2	2
6	**John Surtees**	**McLaren-Ford V8 DFV 950**	1

FRENCH GRAND PRIX – Clermont-Ferrand

1	**Jochen Rindt**	**Lotus-Ford V8 DFV 929**	9
2	**Chris Amon**	**March-Ford V8 DFV 927**	6
3	**Jack Brabham**	**Brabham-Ford V8 DFV 937**	4
4	**Denny Hulme**	**McLaren-Ford V8 DFV 061**	3
5	Henri Pescarolo	Matra-Simca V12	2
6	**Dan Gurney**	**McLaren-Ford V8 DFV 906**	1

BRITISH GRAND PRIX – Brands Hatch

1	**Jochen Rindt**	**Lotus-Ford V8 DFV 901**	9
2	**Jack Brabham**	**Brabham-Ford V8 DFV 937**	6
3	**Denny Hulme**	**McLaren-Ford V8 DFV 061**	4
4	Clay Regazzoni	Ferrari F 12	3
5	**Chris Amon**	**March-Ford V8 DFV 942**	2
6	**Graham Hill**	**Lotus-Ford V8 DFV 056**	1

GERMAN GRAND PRIX – Hockenheim

1	**Jochen Rindt**	**Lotus-Ford V8 DFV 901**	9
2	Jacky Ickx	Ferrari F12	6
3	**Denny Hulme**	**McLaren-Ford V8 DFV 933**	4
4	**Emerson Fittipaldi**	**Lotus-Ford V8 DFV 052**	3
5	**Rolf Stommelen**	**Brabham-Ford V8 DFV 939**	2
6	Henri Pescarolo	Matra-Simca V12	1

AUSTRIAN GRAND PRIX – Osterreichring

1	Jacky Ickx	Ferrari F12	9
2	Clay Regazzoni	Ferrari F12	6
3	**Rolf Stommelen**	**Brabham-Ford V8 DFV 949**	4
4	Pedro Rodriguez	BRM V12	3
5	Jackie Oliver	BRM V12	2
6	Jean-Pierre Beltoise	Matra-Simca V12	1

ITALIAN GRAND PRIX – Monza			Points
1	Clay Regazzoni	Ferrari F12	9
2	**Jackie Stewart**	**March-Ford V8 DFV 057**	**6**
3	Jean-Pierre Beltoise	Matra-Simca V12	4
4	**Denny Hulme**	**McLaren-Ford V8 DFV 061**	**3**
5	**Rolf Stommelen**	**Brabham-Ford V8 DFV 949**	**2**
6	**Francois Cevert**	**March-Ford V8 DFV 053**	**1**

CANADIAN GRAND PRIX – St. Jovite			
1	Jacky Ickx	Ferrari F12	9
2	Clay Regazzoni	Ferrari F12	6
3	**Chris Amon**	**March-Ford V8 DFV 907**	**4**
4	Pedro Rodriguez	BRM V12	3
5	**John Surtees**	**Surtees-Ford V8 DFV 950**	**2**
6	**Peter Gethin**	**McLaren-Ford V8 DFV 902**	**1**

UNITED STATES GRAND PRIX – Watkins Glen			
1	**Emerson Fittipaldi**	**Lotus-Ford V8 DFV 052**	**9**
2	Pedro Rodriguez	BRM V12	6
3	**Reine Wisell**	**Lotus-Ford V8 DFV 901**	**4**
4	Jacky Ickx	Ferrari F12	3
5	**Chris Amon**	**March-Ford V8 DFV 064**	**2**
6	**Derek Bell**	**Surtees-Ford V8 DFV 904**	**1**

MEXICAN GRAND PRIX – Mexico City			
1	Jacky Ickx	Ferrari F12	9
2	Clay Regazzoni	Ferrari F12	6
3	**Denny Hulme**	**McLaren-Ford V8 DFV 061**	**4**
4	**Chris Amon**	**March-Ford V8 DFV 064**	**3**
5	Jean-Pierre Beltoise	Matra-Simca V12	2
6	Pedro Rodriguez	BRM V12	1

1970 World Champion driver	**Jochen Rindt**
1970 World Champion car	**Lotus-Ford V8 DFV**

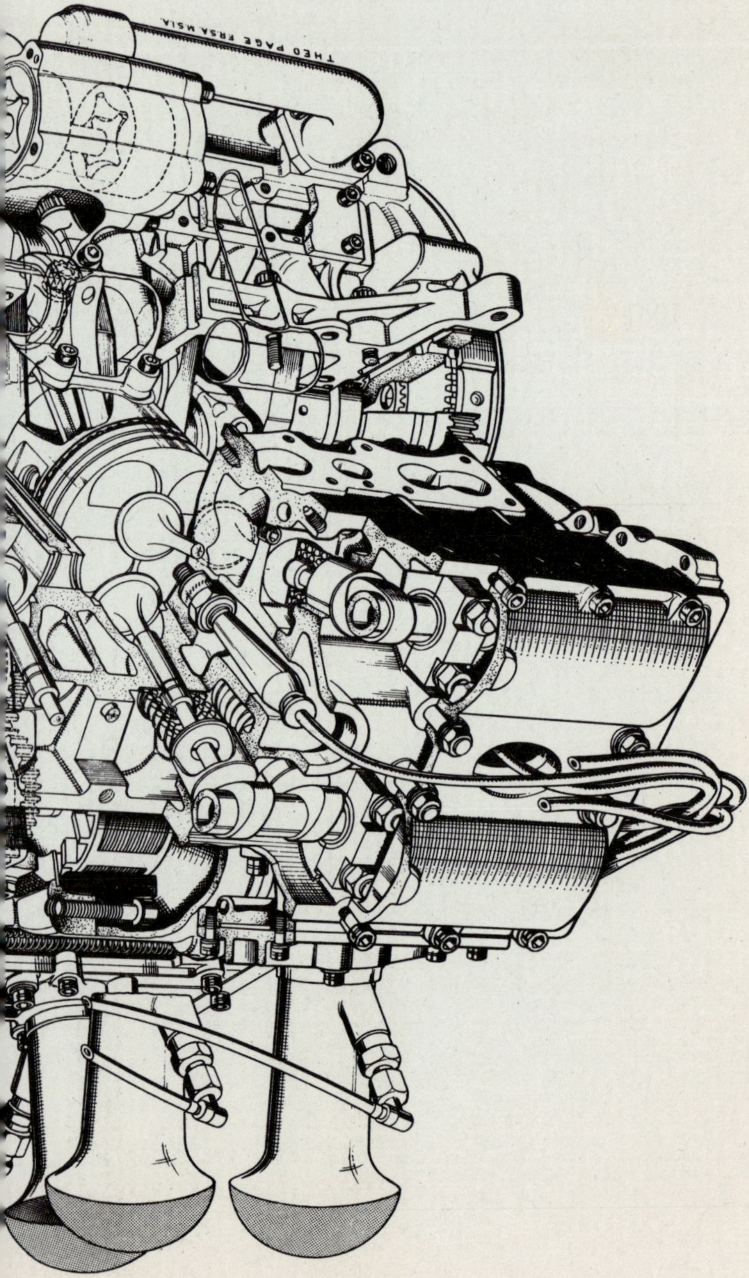
THEO PAGE F.R.S.A. M.S.I.A.